AF477202

Foundations in Pharmaceutical Biotechnology

Foundations in Pharmaceutical Biotechnology

B.P. Nagori

B.Pharm. (Hons.), MBA, M.Pharm., Ph.D., LLB

Professor & Director

Roshan Issarani

M.Pharm., Ph.D.

Professor & HOD, Pharmaceutics

Lachoo Memorial College of Science and Technology, Pharmacy Wing

Jodhpur (Rajasthan)

PharmaMed Press

An imprint of Pharma Book Syndicate

A unit of BSP Books Pvt., Ltd.

4-4-316, Giriraj Lane,
Sultan Bazar, Hyderabad - 500 095.

Published by :

PharmaMed Press
An imprint of Pharma Book Syndicate

A unit of BSP Books Pvt., Ltd.

4-4-316, Giriraj Lane, Sultan Bazar, Hyderabad - 500 095.
Phone: 040-23445605, 23445688; Fax: 91+40-23445611
E-mail: info@pharmamedpress.com
www.pharmamedpress.com/pharmamedpress.net

ISBN : 978-93-85433-18-4 (HB)

A man would do nothing
if he waited until he could do it so well
that no one would find fault
with what he has done.

-Cardinal Newman

Preface

Biotechnology is a broad spectrum, multidisciplinary and an ever changing, vast field. The world over, biotechnology is revolutionizing the development of products, processes and services in healthcare, agriculture, environment and in a range of industries like leather, detergent, textile, food & beverage and cosmetics. Since, the flavour of biotechnology is found in almost every sector, the systematic approach to the study of biotechnology has been specialized sectorwise. Thus, branches like pharmaceutical biotechnology, agricultural biotechnology, environmental biotechnology, leather biotechnology, textile biotechnology, food biotechnology etc. have been introduced in the curriculum. In this book, our aim is to comprehend the principles of biotechnology as applicable to the production of pharmaceuticals.

The field of biotechnology and in particular Pharmaceutical Biotechnology is developing rapidly. Since the introduction of human insulin in 1982, about a hundred biotech based pharmaceuticals have hit the market and several hundred are in the pipeline. Through pharmaceutical biotechnology, it has become possible to produce rare pharmaceuticals and pharmaceuticals that were previously thought impossible to be produced, in sufficient amounts. The novel, biotech-based pharmaceutical products and services/ techniques are being introduced at a very rapid pace than one can imagine. At the onset of this century, Human Genome Project (HGP), the aim of which was to decipher the entire human genetic code, was completed. This has marked the beginning of an era that is already revolutionizing drug development, gene therapy and the entire approach to health care.

Amongst the new drugs being introduced, 40% are based on the principles of pharmaceutical biotechnology. Currently, over 700 biotech based products are under clinical studies. There are about 3000 biotech based companies on the globe, half of them in the US and approximately the same number in Europe especially Britain, having a market of US$ 50 billion. As far as India is concerned, the biotech sector is predicted to grow much faster than the IT sector. The present Indian biotech based market, estimated at US$ 400 million, is set to grow y 40% and the exports by 70% annually. The Pharma industry is all set for a shift towards gene-based drug discovery.

The reader will appreciate the fact that the painstaking efforts of biotechnologists are already turning into reality faster than one can imagine. Biotechnology is merely not restricted to R&D work but it has practical applications with a potential to have a great impact on the mankind. A pharmacist has a long way to go into this century with biotech based products. Future is of biotechnology!

The Indian government, its policy makers and the Pharma sector are aware of the potential of pharmaceutical biotechnology. Thus, formal education in pharmaceutical biotechnology has been felt necessary and has been introduced in universities and colleges at undergraduate and postgraduate levels throughout India. The student finds around him a great deal of literature on pharmaceutical biotechnology. Time in and out, there are breaking news from the world of pharmaceutical biotechnology. The students, and teachers alike, find themselves at times aloof, confused and even puzzled, because they are unable to comprehend and grasp the matter. This is because, pharmaceutical biotechnology is a young discipline and an understanding of the basic underlying principles needs to be developed. Thus, there is a strong need for an introductory textbook on pharmaceutical biotechnology that could lay the foundations on to which the present and the future pharmaceutical biotechnologist could be nurtured.

The fundamental concepts presented in all the chapters shall serve as foundation to the students to understand and interpret the latest research work going on in the field of biotechnology. Topics have been included that not only cover the requirement of syllabi of various universities, but shall also benefit the students in keeping abreast with the latest developments in the field of biotechnology. The book covers the requirement of Biotechnology subject in syllabi as prescribed by AICTE and various Universities for both UG and PG students of Pharmacy, Engineering and Science. The book shall be useful for the students of B.Pharm. & M.Pharm., B. Sci. Biotechnology, M.Sc. Biotechnology and B.E. Biotechnology and the teachers alike.

It is to be mentioned that it is our maiden attempt in the direction of writing a book on pharmaceutical biotechnology which is itself a very young discipline. Attempt has been made to keep the text simple so that the concepts could be made clear. Any suggestions, views and comments from the readers shall be welcome for future improvements.

Prof. Dr. B.P. Nagori

Prof. Dr. R. Issarani

Acknowledgement

It gives us immense pleasure to express our deepest sense of gratitude to Shri N. K. Mathur, Chairman, and Shri R. C. Mathur, Secretary, Lachoo Memorial College Society, for acting as the guiding force and making this commendable task possible.

We must place on record a very special thanks to Ms. Garima Rani Ojha and Ms. Priyanka Sharma, students of Pharmacy, for collection of literature in part.

We would also like to place on note our thanks to one and all who helped us either directly or indirectly in coming out with this book.

Above all, we acknowledge with deep appreciation the indispensable encouragement and moral support of our parents and family members during the course of work.

Prof. Dr. B.P. Nagori

Prof. Dr. R. Issarani

Contents

1 PHARMACEUTICAL BIOTECHNOLOGY : AN INTRODUCTION

1.1   Definition

Biotechnology is defined in different ways by different organizations/ committees of the world. It is also agreed by all that biotechnology cannot be defined in its entirety. However, biotechnology is generally defined as the application of biological organisms, systems, and processes for manufacturing and service industries.

In order to better understand the definition of biotechnology, see Figure 1.1.

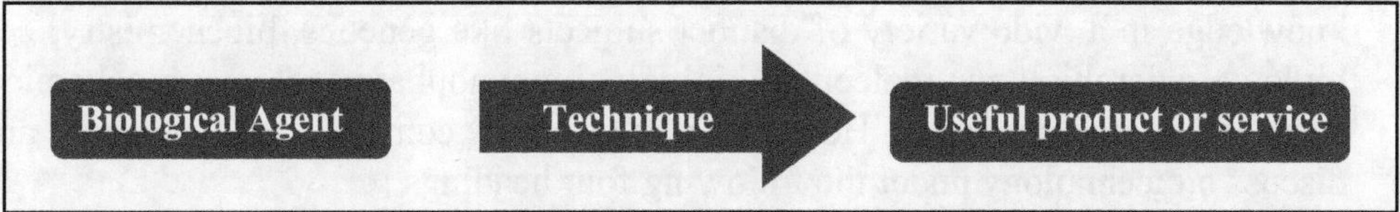

Figure. 1.1 Definition of Biotechnology.

1) **Biological agent :** The biological agent may include (i) Living organisms like microbes, plants, animals or genetically modified (GM) microbes, ; (ii) Substances from these organisms like tissues, cells, enzymes and fragments of DNA.

2) **Technique :** Based on the technique employed, biotechnology is sometimes termed as traditional biotechnology or modern biotechnology :

(i) *Traditional biotechnology :* Under this term, the techniques included are selection and cross breeding used for the improvement of the characteristics like better yields, resistance to diseases, vigor of economically important plants and animals. Also under this is included the technique of fermentation employing microbes to produce beverages (alcohol), foods (idli, bread), antibiotics (penicillin) etc.

(ii) *Modern biotechnology :* In recent years, wherever the term biotechnology is used it generally refers to the sophisticated, novel techniques for genetic manipulation mostly outside the cell. To make it more clear, genetic manipulation comprises of :

- Technique of genetic engineering, wherein a foreign gene (from any biological organism in the universe!) for a desired product is inserted, using vectors, into an organism called the host, so that the gene can be expressed to produce the product in the host.

- Technique of cell fusion, involving formation of a single cell from two different types of cells i.e. the hybrid or cybrid. This is done in order to combine the desirable characteristics of the two cells. For example, a cancerous cell, called myeloma, having the characteristic of immortality and a B lymphocyte, having the characteristic of producing antibodies, are both fused together to give what is called as hybridoma (a hybrid myeloma). This has traits of both the cells and is used for producing monoclonal antibodies. Another example of this technique is protoplast fusion. Protoplast is a plant cell devoid of cell wall. The absence of cellulosic cell wall permits protoplasts of similar or different species/genera to fuse together with advantageous characteristics of both in the hybrid.

3) **Useful product or service :** The goal of employing biological technology i.e. biotechnology (use of biological material and either traditional or modern techniques) is to produce a commercially useful product, modified product (biotransformation) or a service. The examples of such products or services include: humulin, monoclonal antibodies, penicillin, urease, shikonin, alkaloids, steroids, prostaglandins, diagnostics, effluent treatment (bioremediation) etc.

Thus, it can be said that biotechnology is a multidisciplinary field involving knowledge in a wide variety of distinct subjects like genetics, biochemistry, cell biology, microbiology, molecular biology, immunopharmacology, biochemical engineering and biophysics. However, it is generally convenient to categorize and discuss biotechnology under the following four headings :

(i)   Genetic Engineering,

(ii)   Fermentation,

(iii)   Plant/animal cell/tissue culture and

(iv)   Biotransformation.

A pharmacy student, must be aware that WTO proposals have been implemented in India since January 2005. A part of these proposals includes patenting of products. For a product to be patented it should be

(i)   a novelty (new; never known before)

(ii)   an invention (has to be an innovation i.e. made by man and not merely a discovery i.e. for something already existent in nature) and

(iii)   commercially / industrially useful (such that it can be produced at a mass scale to derive economic befits).

Looking at these three criteria, it can be concluded that biotechnology products and services are patentable. Also the biological agent such as a genetically modified (GM) microbe, seed, transgenic plant or animal used for producing the products are also patentable. Here ethical, legal and social issues (ELSI) arise, i.e. whether a life form (a creation of the God) can be patented. However as on today, the genetically modified (GM) life forms are patentable the world over including in India, since, they are an invention of the human mind, making them a private property!

1.2   History

Biotechnology is as old as the origin of first form of life itself. First, the deoxyribonucleic acid (DNA) was formed near the oceans. This over a long period of time got enclosed within a cell wall giving rise to the prokaryote or primitive form of life. This is when foundations of biotechnology were laid down by the nature itself. The genetic material over billions of years underwent recombination and mutation by the laws of nature resulting in diversity in life forms.

Also, the microbes have acted as saprophytes since the origin and have kept the environment clean, generated halometabolites, methane gas etc. The microbes are present in our guts which produce useful substances like vitamin B12, aid in entero-hepatic cycling of bile salts, ferment dietary fibers and produce short chain fatty acids and gases.

But biotechnology as we have defined involves the intervention of human mind and not merely as the things go as desired by nature! Thus a biotechnological process should:

1)   Have an aspect of commercialization i.e. the product or service should generate monetary profits

2)   be able to produce the desired product in huge amounts than that can be produced naturally, so that it is available for use to the masses

3)   be capable of scaling up to the industrial scale

4)   yield a product that is innovative (invention but not a discovery) E.g. , A black or purple rose not red rose!

5)   allow recombination of genetic material not only within a specie but between different species, genera or kingdoms E.g.: Production of human insulin in *Escherichia.coli* or production of human antibodies in plants.

Based on the above account, the history of biotechnology, therefore, dates back to 6000 years B.C that is when man developed methods to use biological agents to achieve commercially and socially desirable goals. This was when man learnt to produce beer using yeast by fermentation process (traditional biotechnology). Later in 4000 B.C., bread was produced using yeast. Starting with all this several thousand years ago, man can now (since the last three decades) manipulate life forms in a purposeful, predetermined manner at the genetic level. Some of the important events in the history of biotechnology are presented in Table 1.1.

1.3   Scope

Biotechnology has a wide ranging scope. It has a role to play, in a lot many industries, directly or indirectly related to our lives. Some of the biotechnology based products and services are listed in Table 1.2.

Since the introduction of human insulin in 1982, over 60 innovative biotech based products treating over 50 diseases have hit the market. A listing of the modern biotech based products is presented in Table 1.3. It is interesting to note that about 75 percent of these products are 'modern' biotechnology products i.e. recombinant DNA products and monoclonal antibodies. Again it is the monoclonal antibodies that top the list of current biotech based research products or the products under clinical trials.

Table 1.1 : Biotech : A Brief history.

Year	Event
6000 BC	Making of wine and beer using yeast
4000 BC	Production of leavened bread using yeast
1670-80 AD	Mining of copper using microbes
1910	Establishment of large scale sewage purification systems
1912-14	Production of industrial chemicals such as acetone, butanol, glycerol
1928	Discovery of penicillin by Alexander Fleming
1944	Production of penicillin at a large scale
1950s	Production of other antibiotics such as tetracyclines, streptomycins
1953	3-dimensional structure of DNA proposed by J.D.Watson & F.H.Crick based on X-ray diffraction of DNA studied by R.E.Franklin & M.H.Wilins
1962	Mining of uranium using microbes in Canada
1962-1971	Site-specific recognition and cleavage of DNA by restriction endonucleases carried out by: W.Arber in 1962; M. Meselson & R.Yuan in 1968; H.O. Smith in 1970; D. Nathans in 1971
1966	Determination of genetic code by M.Nirenberg et al; H.G.Khorana
1967	Identification of DNA ligase by M. Gellert
1970	Identification of RTase by H.M.Temin & S.Mizutani; D.Baltimore
1971-72	Development of DNA cloning techniques by H.W.Boyer, S.Cohen, P.Berg
1973	First successful genetic engineering experiment performed Formal discussions held on emerging rDNA technologies (Gordon conference on Nucleic acids)
1973	Initiation of fuel programme by Brazilian government to replace oil with alcohol
1975	Making of monoclonal antibodies using hybridoma created by C.Milstein & G.Kohler
1976	Introduction of guidelines on genetic engineering by Recombinant Advisory Committee
1977	Discovery of DNA sequencing technologies by F.Sanger & W.Gilbert
1980	Patenting of microorganisms permitted by US court; General Electric was granted patent for 'superbug'
1981	Approval of first MAb based diagnostic kit using anti-CD3 named BioClone by Ortho Diagnostics
1981	Listing of a first biotech company, Cetus, on the stock market
1982	Approval of first ethical pharmaceutical produced using rDNA technology, Humulin, by Genentech & Eli Lilly
1982	Expression of a foreign, bacterial antibiotic resistance gene in plants by Monsanto Co.
1984	Approval of Interferons for protection against cattle diseases
1980 onward	Approval/ production of growth hormone, interferons, new antibiotics by cell fusion, genetically engineered proteins for heart attack & stroke, MAbs against cancer, new vacancies, interferons for cancer
1988	Issue of first patent for genetically engineered mammal, transgenic mouse, to P.Leder & Harvard University
1990 onward	Advancements in DNA & protein sequence & synthesis technologies; Site recent past specific integration of cloned DNA sequences; DNA finger printing; Gene therapy; Embryo transfer & split embryo technology; Animal cloning (Dolly); Human Genome Project; Computer-aided molecular modeling & design

Table 1.2 : Some Biotech based products and services.

Pharmaceuticals based on *Genetic engineering & Hybridoma technology*	Chemicals

Pharmaceuticals based on
Genetic engineering &
Hybridoma technology
(see list of products in Table 1.2)

Fermentation
Penicillin
Cephalosporin
Amylases
Proteases
Vitamins
Amino acids
Dextran

Plant & animal culture
Shikonin
Steroids
Alkaloids
Vaccines

Microbial transformation
Steroids
Prostaglandins

Receptor based drug designing

Gene therapy

Bioprostheses (artificial organs)

Bioassay

Medical/Diagnostics
AIDS detection kits,
Glucose measuring kits

Bioelectronics
Biosensors, biochips

Biofuels
Biogas, ethanol, hydrogen

Chemicals
Ethanol, acetone, butanol, ethylene,
acetaldehyde

Specialty chemicals
Biopolymers, hydroxylated aromatics

Animal agriculture
Veterinary pharmaceuticals;
Development of disease free
seed stock and healthier, higher yielding
animal food

Foods and Beverages
Gluconic acid, amino acids,
baker's yeast, single cell protein,
sweeteners, beer, bread, cheese, Idli
glucose syrup, citric acid, lactic acid,

Plant agriculture
Improved varieties in terms of stress,
herbicide & pesticide resistance;
enhanced ability of photosynthesis,
nitrogen fixation & produce;
seedless varieties; biopesticides

Leather Industry
Tanning of animal skin

Textile & Paper Industry

Environmental Related
Effluent treatment

Mining & Metallurgy
Isolation & purification of metals & minerals

Oil recovery

Table 1.3 : Modern biotech based pharmaceuticals.

Generic name	Brand name	Name of the Company	Therapeutic use
Hormones			
Human Insulin	Humulin	Eli Lilly	Insulin-dependent diabetes
	Novolin	NovoNordisk	
Human Growth	Protropin	Genentech	Growth hormone deficiency
Hormone	Humatrope	Eli Lilly	
	SerostimSerono		
	BioTropin	Bio-tech. General	
Somatotropin	Genotropin	Pharmacia & Upjohn	Short stature in children
	Norditropin	NovoNordisk	
	Nutropin	Genentech	
Follitropin β (FSH)	Follistim	Organon	Ovulatory failure
Follitropin α	Gonal-F	Ares-Serono	Ovulatory failure
Growth Hormone Releasing Hormone	Geref	Serono	GH deficiency in children
Interferons			
Interferon α	Welferon	Glaxo Wellcome	Chronic hepatitis C
Interferon α-2a	Roferon-A	Hoffmann La-Roche	Hairy cell leukemia, AIDS related Kaposi's sarcoma
Interferon α-2b	Intron A	Schering-Plough	Hairy cell leukemia, AIDS related Kaposi's sarcoma
Interferon α-n3	Alferon N	Interferon Sciences	Condylomata acuminate
Interferon γ-1b disease	Actimmune	Genentech	Chronic granulomatous
Interferon β-1b	Betaseron	Berlex Labs	Acute relapsing-remitting MS
Interferon β-1a	Avonex	Biogen	Acute relapsing-remitting MS
Interferon α con-1	Infergen	Amgen	Hepatitis C
Growth factors & Interleukins			
Epoetin α	Epogen	Amgen	Anemias from various causes
	Procrit	Orth Biotech	
Filgrastim (r-metHuG-CSF)	Neupogen	Amgen	Neutropenias
Sargramostim (rh-GM-CSF)	Leukine	Immunex	Myeloid reconstitution
Becaplermin (PDGF)	Regranex	Ortho-McNeil	Diabetic foot ulcer
Aldesleukin (IL-2)	Proleukin	Chiron	Metastatic renal carcinoma, metastatic melanoma
Oprelvekin (IL-11)	Neumega	Genetics Institute	Thrombocytopenia

Table 1.3 Contd...

Generic name	Brand name	Name of Company	Therapeutic uses
Monoclonal Antibodies			
Trastuzumab	Herceptin	Genentech	Metastatic breast cancer
Palivizumab	Synagis	MedImmune/Abbott	Respiratory synctia viral & fatal pneumonia
Infliximab	Remicade	Centocor	Crohn's disease
Muromonab-CD3	Orthoclone OKT 3	OrhtoBiotech	Prenention of organ transplant rejection
Abciximab	ReoPro Centicor		Prevention of blood clots
RituximabRituxan	Genentech		non-Hodgkins lymphoma
Daclizumab	Zenapax Roche		Prevention of kideney transplant rejection
Basiliximab	Simulect	Novartis	Acute organ transplant rejection
Staumomab	OncoScint	Cytogen	Detection, staging & follow up of colorectal & ovarian cancers
Enzymes & Blood factors			
Alteplase	Activase	Genentech	MI, pulmonary embolism, stroke
DNase Dornase α	Pulmozyme	Genentech	Respiratory complications of
Imiglucerase	Cerezyme	Genzyme	cystic fibrosis
Factor VII	Novo-Seven	NovoNordisk	Hemophilia
Factor VIII	KoGENate	Bayer	Hemophilia A
	Recombinate	Baxter	
Factor IX	Benefix	Genetics Inst.	Hemophilia B
Vaccines			
Hepatitis B	Energix-B	SmithKline Beecham	Hepatitis B prophylaxis
	Recombivax HB	Merck	
Miscellaneous			
Hyaluronic acid membrane	Seprafilm	Genzyme	Prevention of adhesions after surgery
Cartilage culturing service	Carticel	Genzyme	Cartilage damage in knees
Skin graft product	Apligraft	Novartis	Wound healing of venous leg ulcers

In the coming years a lot more biotech based products shall be introduced in the markets through out the world. Currently, over 700 biotech based products are under clinical studies. There are about 3000 biotech based companies on the globe, half of them in the US and approximately the same number in Europe especially Britain, having a market of US$ 50 billion. A listing of the world's top ten biotech companies is presented in Table 1.4.

Table 1.4 : World's top ten biotech companies in order of their sales.

1 Amgen	6 Agouron
2 Genentech	7 Centocor
3 Chiron	8 Immunex
4 Genzyme	9 Nabi
5 Biogen	10 MedImmune

As far as India is concerned, the biotech based organizations include Bharat Immunologicals and Biologicals Corporation Ltd. (oral polio vaccine), Indian Institute of Immunology, Hopkins Institute, Shanta Biotech, Biogenie etc. to name a few. Shanta biotech, a Hyderabad based company, has products such as Shanvac, the indigenously developed hepatitis vaccine; Shanferon, a preparation containing interferon etc.

As per the report of associated chambers of commerce (Assocham), the biotech sector is predicted to grow much faster than the IT sector. The present Indian biotech based market, estimated at US$400 million, is set to grow by 40% and the exports by 70% annually. The Pharma industry is all set for a shift towards gene-based drug discovery. An account of the biotech industrial sector is depicted in Figure 1.2.

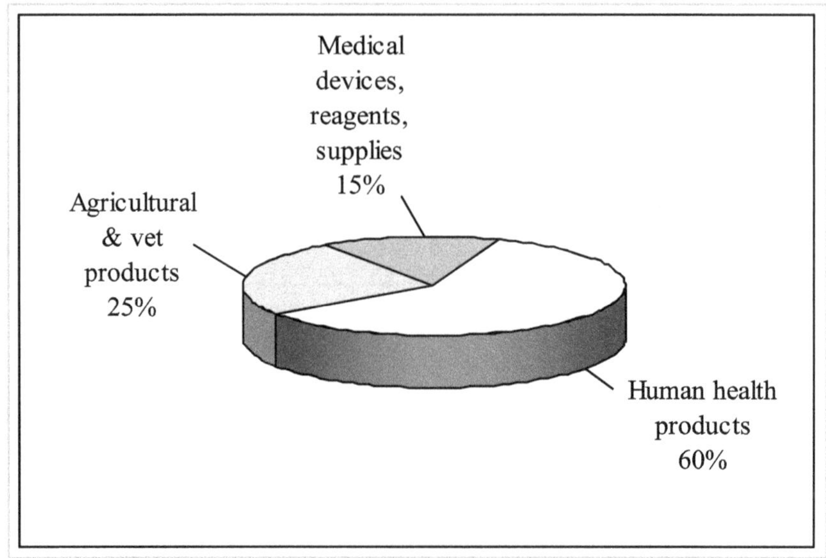

Figure. 1.2 : An account of Biotech Industry.

So, one has the scope to join, as a biotechnologist, in any of the biotech industrial sectors. Further, one can join academics and 'cultivate' well trained biotechnologists. Also one can provide consultancy services to the practicing pharmacist. But one must have a basic knowledge of all the subjects mentioned earlier and should have an in depth knowledge of at least one discipline so that one can make significant contribution to the team of biotechnologists.

1.4   Future Perspectives

It is rightly said that present is of Information Technology future shall be of Biotechnology. Take for example, if you conceive an idea of a purple rose, than you can create a picture of the rose or take a snap of the natural red colored rose, put it into the computer and fill purple color into the petals. But with biotechnology, if you conceive an idea of purple rose, than you shall have to manipulate genes of the naturally occurring red rose and produce ation not a discovery) a purple rose. This is to say that, with information technology you can have things created in 'virtual reality' (cyber space) but with biotechnology you have things produced in 'real'. However, sometimes it is better to have things in virtual reality as for instance in case of testing of a nuclear weapon. Also, if you are creating an animation picture and if you happen to create a deadly, dangerous monster than you always have the option of editing and can 'undo'. But with biotechnology, once the genes are manipulated and the deadly dangerous monster is produced, it will be too difficult, if not impossible, to contain its production (replication). It is to be mentioned here that information technology and biotechnology revolutions are not likely to continue as separate phenomena. Rather biotechnology and computing will interact increasingly with each other and shall contribute to the pace of advance in each field. Already a new field called bioinformatics has emerged in the last decade as a result of the marriage between biology and computer science. Bioinformatics is defined as the science of developing computer databases and algorithms, managing data and mining data for the purpose of speeding up and enhancing biological research especially in the field of genomics and proteomics. Therefore, biotechnology, information technology and bioinformatics have the potential to change the face of life on earth.

Keeping aside the risks involved in using biotechnology and despite the fact that there are controversies already arising at the international level on the use of genetically modified (GM) seeds, plants, animals and microbes and the consumer/social groups debating on the ethical, legal and social issues (ELSI), biotechnology holds truly a promising future.

1.5   Message for the Pharmacists

Biotechnology based pharmaceuticals which are to be used for prevention, treatment, diagnosis or altering the physiological functions include: antibiotics, enzymes, genetically engineered vaccines, monoclonal antibodies, interferons, growth hormones, insulin, alkaloids, steroids etc.

Now, as a pharmacist, who is a competent authority on drugs, you should have thorough knowledge of biotechnology based products. First of all, the pharmacokinetics i.e., liberation, absorption, distribution, metabolism, excretion and toxicity (LADMET) of the biotechnology based products should be known. Next, their pharmacodynamics i.e., mechanism of action has to be known. Further a pharmacist should have knowledge about the production techniques, storage, packaging, expiration etc. of these products.

Just like the conventional drugs, biotechnological products, also have drug-drug, drug-food and drug- disease interactions. It is again the responsibility of the practicing pharmacist to keep himself abreast on the knowledge concerning the clinical use of biotechnology based products including their therapeutic use and drug monitoring, so as to provide safe and effective treatment. Responsibility of including drugs in the hospital formulary lies on the shoulders of the hospital pharmacist. Before inclusion of biotechnology based products in the formulary, one has to exercise precaution and use sound judgment. The pharmacist should judge the efficacy of biotechnology based products in comparison to the conventional drugs available for treatment and than decide in their favor.

Most of the modern biotechnology based products are genetically engineered products. And a gene product as you know is always a protein or a peptide. Thus, one important area that has emerged is the invention of novel drug delivery systems for these biotechnology based products for which oral route of administration is unsuitable.

Last, but not the least, a professional pharmacist should be a computer literate and must have access to internet facility. This is because, biotechnology is a rapidly and a globally developing field, a number of biotechnological products and services are being introduced and the knowledge base is continually expanding, so it is only through the 'world wide web' (www), a practicing pharmacist can refresh and update his knowledge.

Wishing all the Pharmacists a happy, prosperous and a healthy journey!

2

GENETIC MATERIAL AND ITS ENGINEERING

CHAPTER CONTENTS AT A GLANCE

2.1   Introduction

We have seen in chapter one that modern biotechnology involves manipulation of genetic material i.e. insertion of foreign genetic information into fast growing microorganisms to produce a product having a high yield and a high rate of production. The underlying principle in any of the four techniques of pharmaceutical biotechnology i.e. genetic engineering, fermentation, plant/animal cell/tissue culture or biotransformation, is that the product that the biological agent produces depends on the genetic information contained in it. Thus gene is the foundation stone of biotechnology. Enzymes, the biocatalysts, in the presence of which all the biotechnology based products are produced and using which (restriction endonucleases) genetic engineering is made possible are also the products of genes.

The central dogma of life which is at the heart of biotechnology is depicted in Figure 2.1.

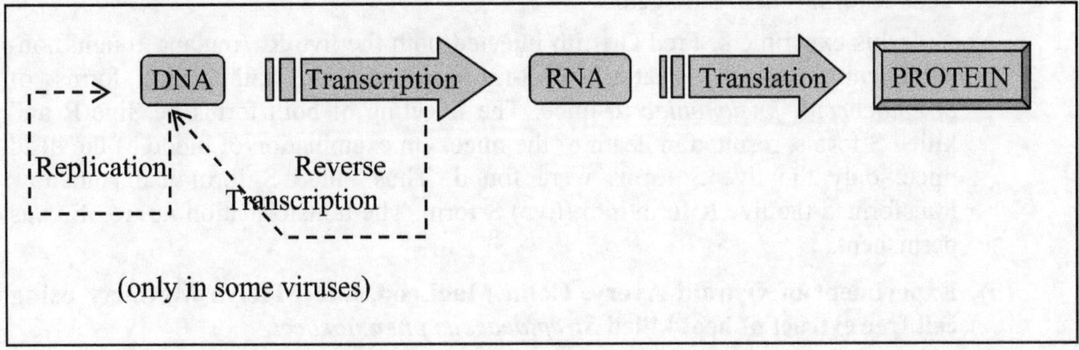

Figure. 2.1 : Central dogma of life.

The central dogma of molecular biology applies to all organisms in the universe, since the genetic language is universal. DNA is the starting material that serves as template for its replication and the information coded in the genes is transcribed into RNA and translated into protein i.e. gene products.

Thus it is appropriate to first talk about genes and their manipulation i.e. genetic engineering in this chapter and then talk about enzymes and other biotech based products in the subsequent chapters.

2.2   DNA is the Molecule of Heredity

Prior to mid twentieth century, proteins were believed to carry genetic information and were considered to be the molecules of heredity. However, scientists at that time postulated that for a molecule to carry genetic information, it should:

1. be stable,
2. replicate itself,
3. have a high degree of fidelity i.e. there should be no loss or alteration of genetic information as it is being transferred from one generation to next over a period of time and
4. have flexibility so that it can account for diversity and evolution of life forms.

Thus in the early part of the last (20th) century, the widely held notion that proteins are the molecules of heredity was challenged. Scientists performed experiments to prove that DNA is the molecule that fulfills all the above mentioned requirements and is actually the molecule of heredity. An account of three such noted experiments is presented below.

(i)   **Experiment of Fred Griffith using R&S forms of *Streptococcus pneumoniae***

Fred Griffith, in the year 1928, performed an experiment using R&S forms of *Streptococcus pneumoniae*. The S form is normal and is pathogenic which causes pneumonia in humans and in other susceptible mammals. It is referred to as the S form because it forms smooth colonies on culturing. The R form of *Streptococcus. pneumoniae* is the mutant form. It is referred to as R form since this form produces rough colonies in culture. The R mutant form lacks an enzyme needed for the synthesis of the capsule which helps to evade detection by the immune system. Thus R form is non-pathogenic.

In his experiment, Fred Griffith injected both the live R- (mutant, rough, non-pathogenic) and the heat-killed S-(normal, smooth, pathogenic) forms of *Streptococcus. pneumoniae* to mice. The injecting of both forms i.e. live R and killed S forms, resulted in death of the mice. On examination of blood of the dead mice, only the live S forms were found. Thus killed S form had somehow transformed the live R form into (live) S form. The transformation i.e. R→S, was permanent.

(ii)   Experiment of Oswald Avery, Colin MacLeod, and Maclyn McCarty using cell free extract of heat killed *Streptococcus pneumococci*.

In 1944, Oswald Avery, Colin MacLeod, and Maclyn McCarty carried out an experiment and found that the R→S transformation can be brought about even under *invitro* conditions.

They added a cell-free extract of heat-killed *Streptococcus pneumococci* in a growing culture of the R form and found that R→S transformation had occurred. Further, they fractionated the extract and assayed its components. They made the following observations:

(a)   Elemental chemical analysis of the active transforming principle was carried out which agreed with that for DNA,

(b)   The optical, ultra centrifugal, diffusive and electrophoretic properties of the active transforming material were like DNA,

(c)   Extraction and removal of protein or lipid fractions from the extract did not result in the loss of transforming ability of the cell-free extract,

(d)   Addition of trypsin, a polypeptide cleaving enzyme, did not result in the loss of transforming ability of the cell-free extract,

(e)   Addition of ribonuclease, an RNA cleaving enzyme, did not result in the loss of transforming ability of the cell-free extract and

(f)   Addition of deoxy-ribonuclease, a DNA cleaving enzyme, resulted in the loss of transforming activity.

This experiment was a landmark and proved that genes are made of DNA i.e. DNA has genetic specificity and that DNA is the 'heredity' molecule.

(iii) Experiment of Alfred Hershey and Martha Chase using radiolabeled bacteriophage and *Escherichia coli*.

The experiment of Alfred Hershey and Martha Chase was based on transduction, i.e. transfer of genetic material from a bacteriophage (a virus that infects bacteria) to bacteria.

Bacteriophage i.e. virus, consists of DNA core encapsulated in a protein coat. In their experiment, carried out in 1952, Alfred Hershey and Martha Chase radiolabeled the phage DNA with ^{32}P, while the protein coat was radiolabeled with ^{35}S. Next, they infected a culture of *Escheichia.coli* with the radiolabeled phage. After a short incubation period, the culture was centrifuged. This resulted in the formation of a pellet (consisting of bacteria) at the bottom of the tube and supernatant containing small particles (protiens). These two fractions were analyzed for ^{32}P and ^{35}S to determine the location of phage DNA and protein coat. The phage DNA was found in the infected bacteria in the pellet, while the phage protein was found in the supernatant.

This experiment also proved that a physical separation of the phage into genetic and non-genetic parts is possible. Thus this experiment corroborated the findings of earlier experiments to prove and establish the genetic role of DNA.

2.3   Genome

Genome is the complete genetic content in a cell of an organism. It is the entire DNA in a cell of an organism including its genes and a lot of DNA that does not contribute to genes. Each organism has its own unique genome.

To make the picture clearer, consider one cell among trillions of cells that make up a human. Inside the human cell lies a nucleus. The nucleus contains 23 pairs of chromosomes. One chromosome of every pair is inherited from each parent. An individual human chromosome is made up largely of a tightly coiled double stranded DNA, apart from proteins like histones. At a basic level, thus, the genome is DNA, a natural polymer built up of repeating nucleotides, each consisting of a single sugar, a phosphate group, and one of four nitrogenous bases. Only 2 percent of this DNA, i.e. of human genome, constitutes genes which code for proteins, the remainder 98 percent constitutes the non-coding regions, referred to as 'junk DNA'. The function of junk DNA may possibly be to provide chromosomal structural integrity and to regulate as to what, when, where and how much of a protein is made; for instance compare the growth of hair on the head, arms, and legs while on the soles it is absent!

A gene is a discrete linear sequence of DNA which corresponds to a heritable trait. Genes determine hereditary traits since they carry the instructions, which lead to brown, black or blonde hair, or blue, black or green eyes, or determine whether we are susceptible to diseases such as diabetes and cancer, or even dictate whether an organism is human or another species such as yeast, rice or fly, all of which have their own genomes.

Thus to better understand ourselves biologically, our evolution, our existence and interrelationship with other organisms, scientists are trying to decipher (work out) the genomes of various organisms. It is indeed a Herculean task to determine the genome of millions of species, human being one such species. A brief account of the human genome project is presented below.

The U.S. Human Genome Project (HGP), which began in Oct. 1990, was a coordinated effort of the Department of Energy (DOE) and the National Institute of Health (NIH). The goals of project were to:

1. identify all the approximate 30,000 genes in human DNA,

2. determine the sequences of the three billion chemical base pairs that make up a human DNA,

3. store this information in databases,

4. improve tools for data analysis,

5. transfer related technologies to the private sector, and

6. address the ethical, legal and social issues that may arise from the project.

It was initially planned to last for fifteen years, but it is to be noted that ideas in the field of biotechnology are turning into reality much faster than one thinks. Thus, in February 2001, a working draft of human genome DNA sequence was published by HGP and separately by Celera Genomics, a private company. Due to rapid technological advances the project took thirteen years instead of fifteen years and was completed in April 2003. Coincidentally, the year also marks the fiftieth anniversary of Watson and Crick's publication of DNA structure.

The findings of HGP put forward the fact that we are made of just 30,000 genes only. This newly estimated number of genes is only one third as great as previously believed. Chromosome one has the maximum number of genes (2968) while chromosome Y has the least number of genes (231). On an average a gene consists of 3000 bases, but sizes vary greatly, while a chromosome ranges in size from 50 million to 250 million bases. Each one of our cells, i.e. the human genome, contains about six billion base pairs of DNA, three billion or to be more precise 3164.7 million inherited from each parent. Almost all (99.9%) the nucleotide bases are exactly the same in all people. Our differences are determined by only about one base-pair in each thousand. Scientists have already identified such 1.4 million locations where single –base DNA differences occur in humans.

Having talked about the human genome, a microscopic enormity, we shall now consider the structure, properties and replication of DNA at its very basic: the molecular level.

2.4   Structure of DNA

Deoxyribonucleic acid or DNA is a very long, thread like macro-molecule made up of a large number of deoxyribonucleotides. A nucleotide consists of a nitrogenous base, a sugar and one or more phosphate groups. The sugar moiety is always a pentose. Deoxy points to the fact that the pentose sugar (ribose) lacks an oxygen atom. The nitrogenous bases are of two types-purines and pyrimidines. Two purines: adenine and guanine, and two pyrimidines: cytosine and either thymine or uracil are present in nucleic acids. There are two different types of nucleic acids. One contains the sugar, ribose and is therefore called ribonucleic acid or RNA. The other contains a slightly different sugar called deoxyribose and is therefore called deoxyribonucleic acid or DNA. RNA is found mainly in the nucleolus (a structure within the nucleus) and the cytoplasm, and to a very little extent in the chromosomes. DNA on the other hand is found mainly in the chromosomes. Both types of nucleic acids contain cytosine and have the same purine bases, but whereas thymine occurs only in DNA, uracil occurs only in RNA. The basic structural units of the nucleic acids are shown in Figure 2.2.

Now, if genes are composed of DNA it is necessary that the latter should have a structure sufficiently versatile to account for the great variety of different genes and yet at the same time be able to reproduce itself in such a manner that an identical replica of itself is formed at each cell division. In 1953, based on X-ray diffraction studies, M.H.F.Wilkins, F.H.C.Crick, and J.D.Watson proposed a structure for the DNA molecule which fulfilled all the essential requirements. For their work, Wilkins, Crick and Watson were awarded the Nobel Prize for medicine and physiology.

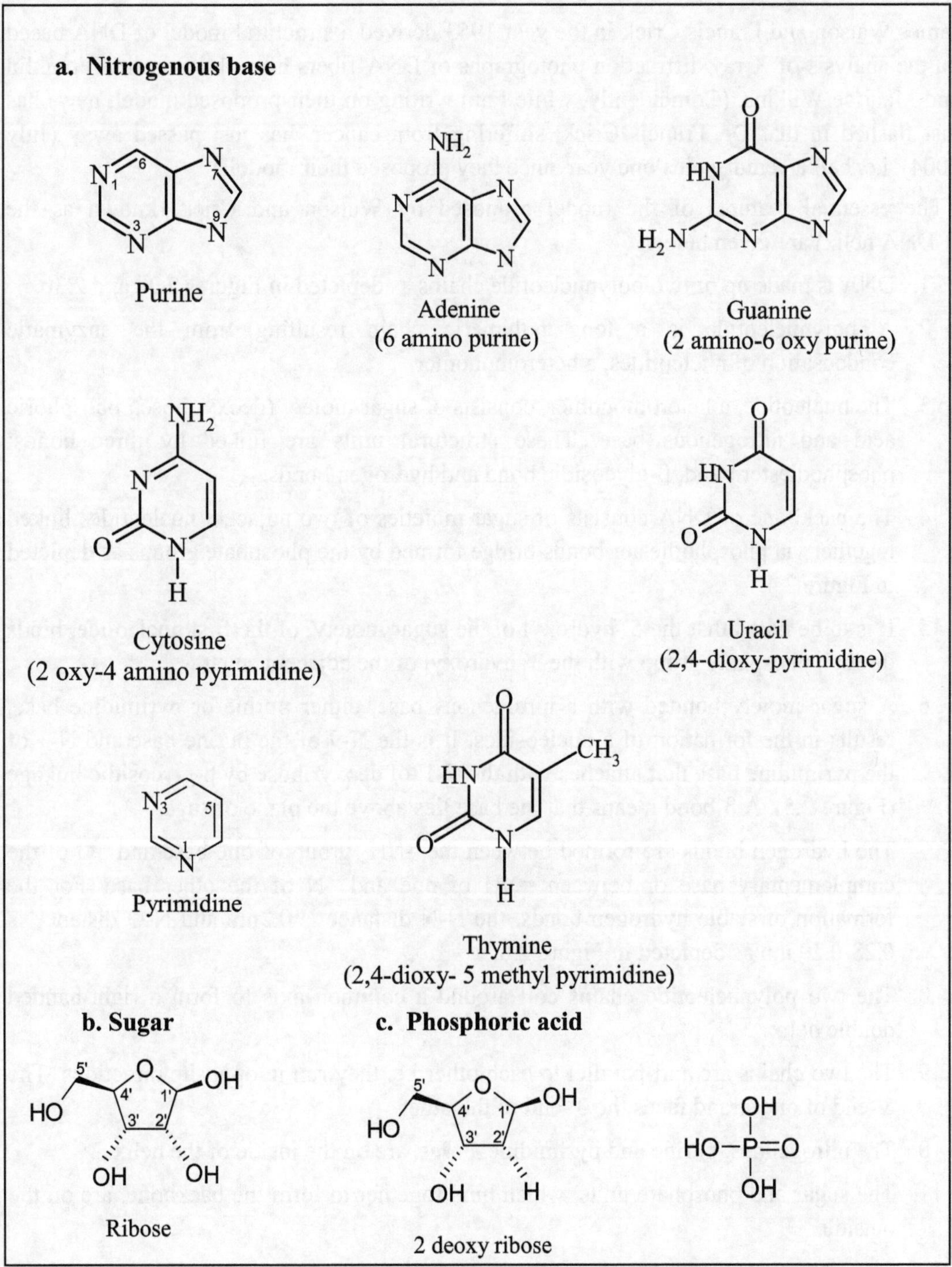

Figure 2.2 Basic nucleic acid structure.

2.4.1 Watson and Crick's model of double helical DNA.

James Watson and Francis Crick in the year 1953 derived a structural model of DNA based on the analysis of X-ray diffraction photographs of DNA-fibers taken by Rosalind Franklin and Maurice Wilkins. (Coincidently, while I am writing on their proposed model, news has just flashed in that Dr. Francis Crick, suffering from cancer, has just passed away (July 2004), i.e. half a century plus one year since they proposed their model!)

The essential features of the model proposed by Watson and Crick (known as the B-DNA helix) are given below:

1. DNA is made up of two polynucleotide chains as depicted in Figures 2.3a and 2.3b.

2. A polynucleotide is a long polymeric chain resulting from the enzymatic condensation of nucleotides, a heteromonomer.

3. The nucleotide, a heteromonomer, consists of sugar moiety (deoxyribose., phosphoric acid and nitrogenous base. These structural units are linked by three bonds: phosphodiester bond, β-glycosidic bond and hydrogen bonds.

4. The backbone of DNA consists of sugar moieties of two adjacent nucleotides linked together via phosphodiester bonds/bridge formed by the phosphate groups as depicted in Figure 2.4.

5. It is to be noted that the 5′-hydroxyl of the sugar moiety, of the first nucleotide, binds through phosphate group with the 3′-hydroxyl of the adjacent sugar.

6. A sugar moiety bonded with a nitrogenous base, either purine or pyrimidine base, results in the formation of a nucleosides. It is the N-9 of the purine base and N-1 of the pyrimidine base that attaches with the C-1′ of deoxyribose by β-glycosidic linkage (Figure 2.5). A β-bond means that the base lies above the plane of sugar.

7. The hydrogen bonds are formed between the $-NH_2$ group of one base and $=O$ of the complementary base or between $=NH$ of one and $-N$ of the other base. For the formation of stable hydrogen bonds, the N-N distance is 0.3nm and N-O distance is 0.28-0.29 nm as depicted in Figure 2.6.

8. The two polynucleotide chains coil around a common axis to form a right-handed double helix.

9. The two chains are anti-parallel to each other i.e. they run in opposite directions. The 5′-end of one strand faces the 3′-end of the other.

10. The nitrogenous, purine and pyrimidine, bases, are on the inside of the helix.

11. The sugar and phosphate units, which bind together to form the backbone, are on the outside.

12. The nitrogenous bases are perpendicular to the helix axis.

13. The planes of nitrogenous bases and sugars are at right angles to each other (a β-bond).

14. The double helix has a diameter of 20 $\overset{\circ}{A}$ i.e. 2nm.

15. One complete turn of the helix, which has a length of 34 $\overset{\circ}{A}$ is referred to as helical pitch.

16. There are 10 bases/turns in B-DNA. This means adjacent bases are separated by a distance of 3.4 $\overset{\circ}{A}$.

17. The bases are stacked one top of the other, related to the next along the chain, by a rotation of 36°. To make this point clear, visualize a spiral staircase and assume that the DNA looks like a spiral staircase. Now from the top floor, look down through the staircase. The case is like the sugar-phosphate backbone, while the stairs are like nitrogenous bases. Each step is related to the next step by some degree of rotation, since the staircase is spiral. One complete turn means 360°. Suppose that in one turn there are ten steps then the adjacent steps are going to be related through an angle of 36°.

18. It is the hydrogen bonds between the pairs of nitrogenous bases of the two strands that hold the chains together.

19. Adenine, a purine, always pairs with thymine, a pyrimidine, by two hydrogen bonds (A=T). Likewise, guanine, another purine, always pairs with cytosine, another pyrimidine, by three hydrogen bonds (G≡C).

20. The sequence of bases along the length of the chain is not restricted in any way. In-fact, it is the precise sequence of bases that carries the genetic information.

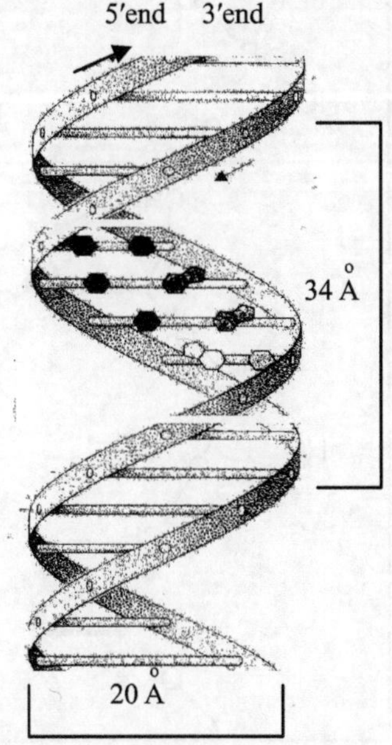

Figure. 2.3 a : Molecular Structure of DNA: front view.

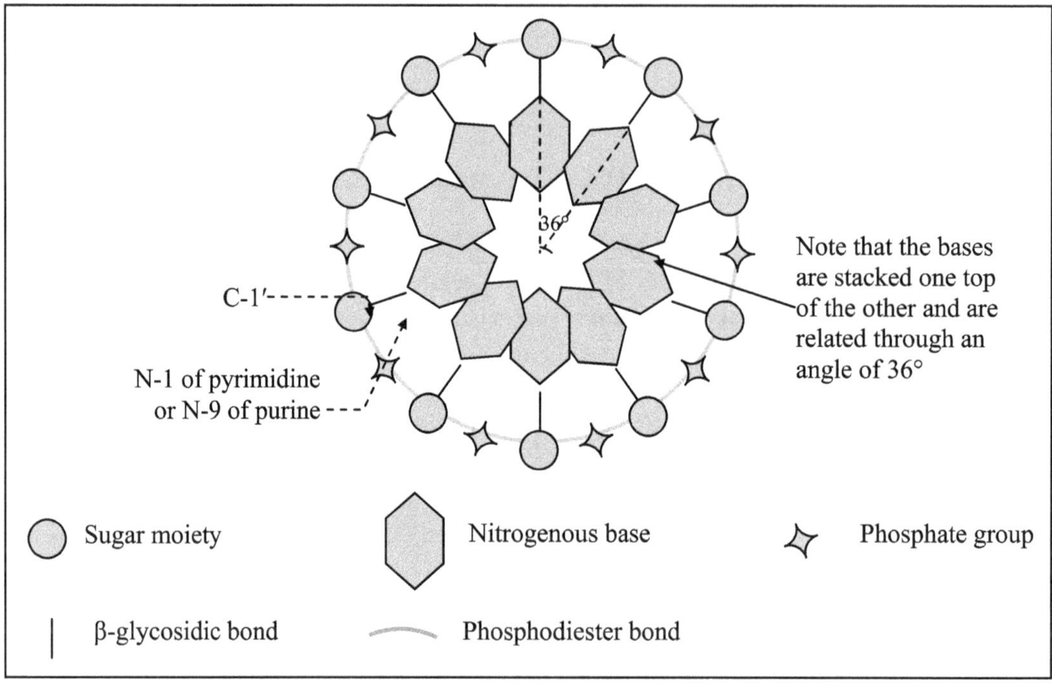

Figure. 2.3 b : Artistic imagination of the top view of
DNA as seen through the helical pitch.

Only one strand is shown for simplicity.

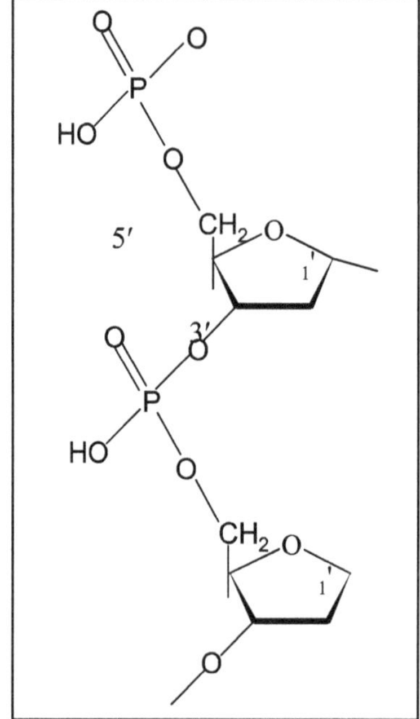

Figure. 2.4 : Sugar-Phosphate bond in DNA backbone.

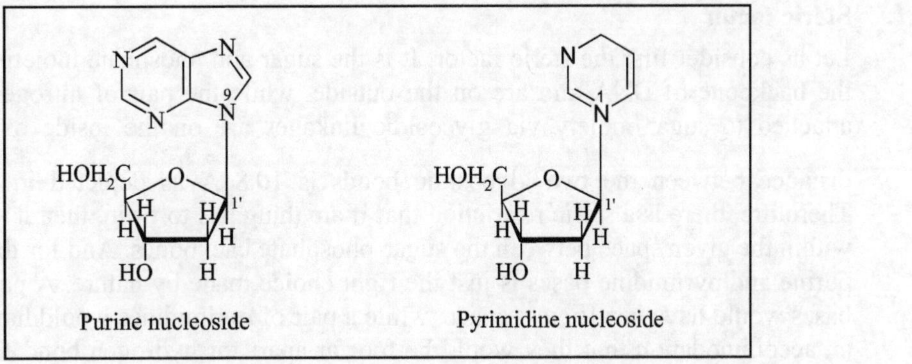

Figure. 2.5 : Nucleoside formation : note the linkage between sugar and nitrogenous base : C1′ with N9 of purine and C1′ with N1 of pyrimidine.

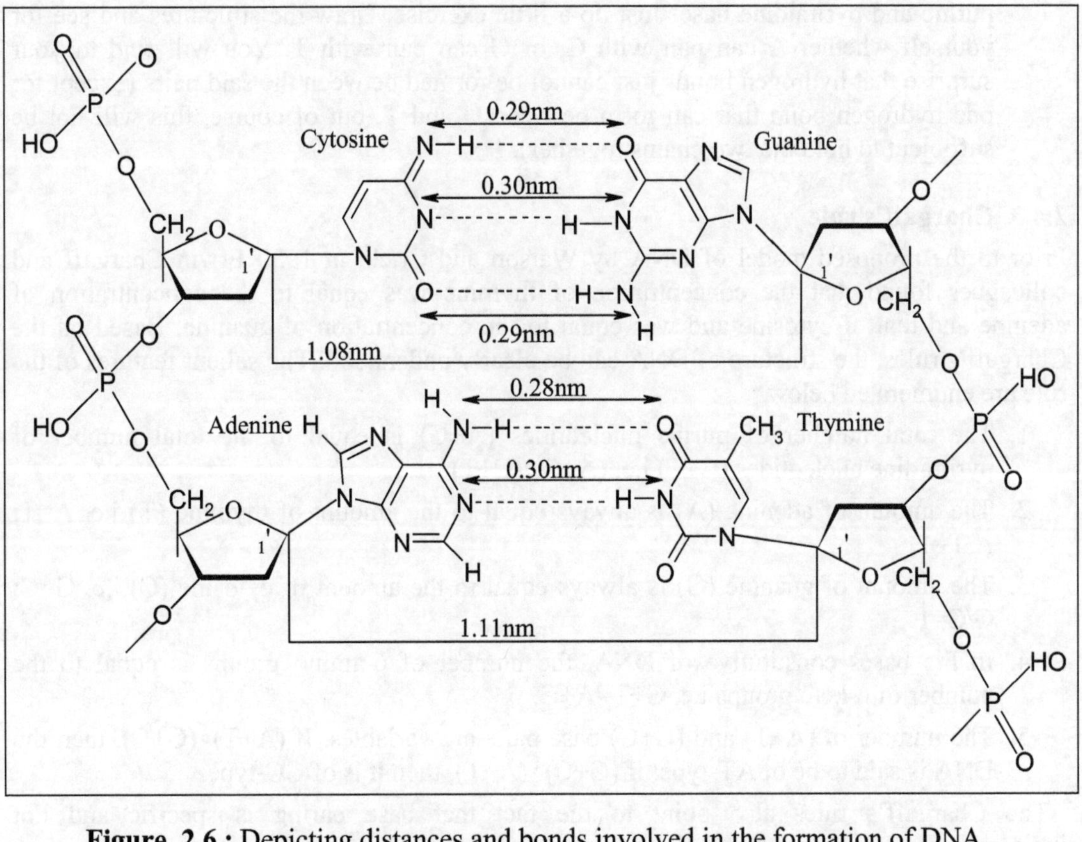

Figure. 2.6 : Depicting distances and bonds involved in the formation of DNA.

2.4.2 Why Guanine always pairs with Cytosine and Adenine pairs with Thymine

Watson and Crick put forward two factors due to which they deduced that guanine must always pair with cytosine and adenine with thymine. It is the steric and hydrogen bonding factors which are responsible for the specificity of pairing of bases.

1. **Steric factor**

    Let us consider first the steric factor. It is the sugar and phosphate moieties that form the backbone of DNA and are on the outside, while the pair of nitrogenous bases, attached to sugar moiety via glycosidic linkages are on the inside of helix. The

    distance between the two glycosidic bonds is 10.8 $\overset{o}{A}$ as depicted in Figure 2.5. Therefore, there is a steric restriction that if anything has to fit in, then it should fit in within the given space between the sugar-phosphate backbones. And for this, a pair of purine and pyrimidine bases is just the right choice made by nature. A pair of purine bases would have insufficient space, while a pair of pyrimidines would have too large an accommodation that they would be too far apart for hydrogen bond formation to occur (for which the optimum distance is 0.28-0.3 nm).

2. **Hydrogen-Bonding factor**

    Next let us consider the factor concerning the formation of hydrogen bonds between a purine and pyrimidine base. Just do a little exercise. Draw the structures and see for yourself whether A can pair with C; or G can pair with T. You will find to your surprise that hydrogen bonds just cannot be formed between the said pairs (except for one hydrogen bond that can form between G and T, but of course, this will not be sufficient to hold the two chains together).

2.4.3 Chargaff's rule

Prior to the proposed model of DNA by Watson and Crick, in 1950 Erwin Chargaff and colleagues found that the concentration of thymine was equal to the concentration of adenine and that of cytosine and was equal to the concentration of guanine. Based on the Chargaff's rules, the structure of DNA can be clearly understood. The salient features of the rule are enumerated below.

1. The total number of purine nucleotides (A+G) is equal to the total number of pyrimidine nucleotides (C+T) i.e. (A+G)/(C+T) =1.
2. The amount of adenine (A) is always equal to the amount of thymine (T) i.e. A=T; A/T=1.
3. The amount of guanine (G) is always equal to the amount of cytosine (C) i.e. G=C; G/C=1.
4. In the bases constitutive of DNA, the number of 6-amino groups is equal to the number of 6-keto groups i.e. G+T=A+C.
5. The number of (A+T) and (G+C) base pairs are variables. If (A+T)>(G+C), then the DNA is said to be of AT-type; if (G+C)>(A+T), then it is of GC-type.

The Chargaff's rules also point to the fact that base paring is specific and not indiscriminate.

2.4.4 Alternate forms of DNA

The three dimensional structure of DNA described as per Watson and Crick's model, is known as the B-DNA. Alternate helical types of DNA also exist, though represented in minority. The two alternate types are the A-DNA and Z-DNA. These are compared in Table 2.1.

Table 2.1 : Alternate helical forms of DNA: A comparison

Point of comparison	Helix type		
	A-	B-	Z-
Shape Broadest	Intermediate	Narrowest	
Rise per base pair	2.3 Å	3.4 Å	3.8 Å
Helix diameter	25.5 Å	23.7 Å	18.4 Å
Screw sense	Right-handed	Right-handed	Left-handed
Glycosidic bond	*anti*	*anti*	*anti* for C,T *syn* for G
Base pairs per turn of helix	11	10	12
Pitch per turn of helix	25.3 Å	35.4 Å	45.6 Å
Tilt of base pairs from normal to helix axis	+19°	$-1°$	$-9°$
Major groove	Narrow and very deep	Wide and deep	Flat ie. absent
Minor groove	Very broad and shallow	Narrow and deep	Very narrow and deep
Other remarks	*a*	—	*z*

a-Closely resembles double stranded RNA; presence demonstrated *in vitro* in less hydrous environment and high Na^+ & K^+ concentration

z- Presence demonstrated *in vitro* under either high salt concentration or in the presence of spermine & spermidine; high degree of negative supercoiling; high degree of methylation at C5

2.4.5 Recent insights into the structure of DNA

Watson and Crick had deduced the structure of DNA based on the X-ray diffraction pattern of DNA fibers. Richard Dickerson and his co-workers obtained much more information on the structure of DNA, based on X-ray analysis of crystallized DNA dodecamer. Thus we now have new insights into the structure of DNA, which include the following:

1. *Angle of rotation* is *28°-42°:* The bases in a chain are related to the next, along the chain, by a rotation varying between *28°-42°* and not rigidly by a fixed rotation of 36° as per Watson and Crick model. Rotation of 28° corresponds to less tight winding and a rotation of 42° corresponds to more tight winding in comparison to the winding that takes place in the chain, when residues (bases) are related by a rotation angle of 36°.

2. *Propeller twist :* The purine and pyrimidine bases are twisted with each other and are not co-planar as believed earlier (see Figure 2.7). The twist of bases is analogous to the blades of propeller; hence it is referred to as propeller twist. This results in stacking of bases along the chain.

3. *Arcing and supercoiling :* The impression one gets by seeing the Watson and Crick model of DNA, is that the entire DNA molecule is linear. If it is considered to be linear, then it will simply not fit into a cell. The DNA can in fact be bent into an arc and can also be supercoiled. The point to be noted here is that, the bending and supercoiling of DNA (so that it can be packaged into chromosomes in a cell) is achieved in an easy way so that large changes in local structure are not brought about.

4. *Kinking :* Further, DNA can be bent (i.e. not just arc) at discrete sites (such as adenine residues). This is referred to as kinking.

5. *Major and minor grooves :* The two glycosidic bonds (between sugar and the base) in the chain are not diametrically opposite to each other. Because of this, DNA has a major groove which is 12 $\overset{o}{A}$ wide and a minor groove which is 6 $\overset{o}{A}$ wide (see Figure 2.8). The hydrogen bonding possibilities in the major and minor grooves are presented in Table-2.2. Because of its larger size, it is the major groove that takes part in interactions with proteins.

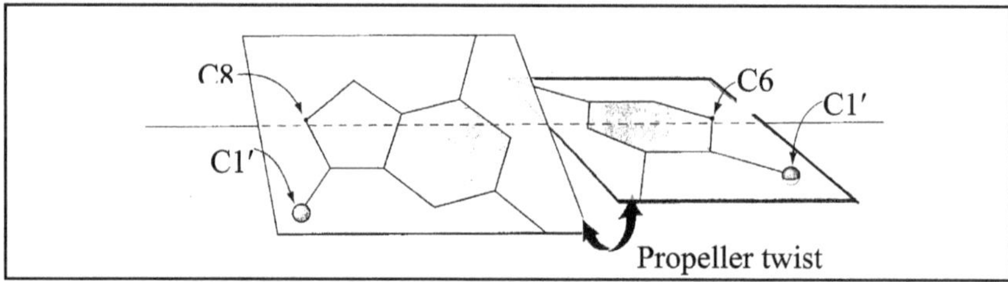

Figure. 2.7 : Depicting propeller twist.

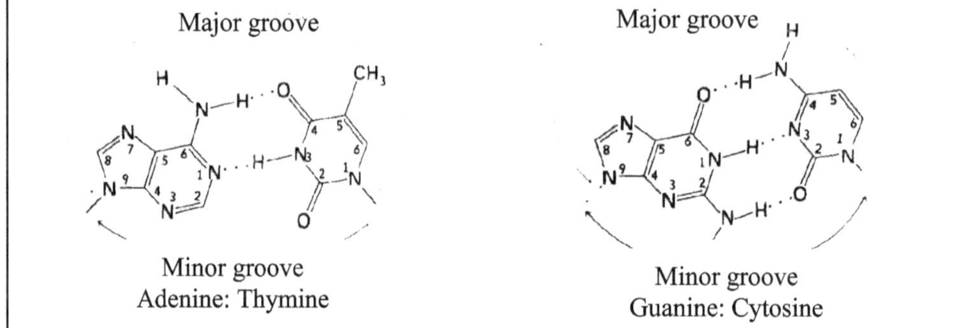

Figure. 2.8 : Depicting major and minor grooves.

Table 2.2 : Hydrogen donors and acceptors in major and minor grooves

Groove	Hydrogen donor	Hydrogen acceptor
Major	Amino group at C-6 of A	N-7 of A&G
	Amino group at C-4 of C	O-4 of T
		O-6 of G
Minor	Amino group at C-2 of G	N-3 of A&G
		O-2 of T&C

2.5　Properties of DNA

The properties of DNA are summarized below in point form.

1. DNA dissolves in water and in many other solvents, producing a viscous solution.

2. Its solution gives acidic reaction.

3. It has absorption maxima at 260 nm (in *u.v.* range).

4. The ratio of absorption at 260 nm by 280 nm is = 2 for pure DNA.

5. Absorption =1 at 260 nm for a DNA concentration of 50 μg/ml.

6. It is denatured i.e. the hydrogen bonds between the complementary bases break open, by heating, treatment with acid, alkali, urea, formamide under *in vitro* conditions. While it is denatured *in vivo* by helix destabilizing, proteins for example e.g. during replication. The opening up of the two strands is referred to as the melting of DNA. The acid or alcohol acts by changing the ionic properties of the groups involved in hydrogen bond formation.

7. Melting of DNA, i.e. falling apart of the two strands, is associated with hyper chromaticity. Hyper chromaticity means increase in absorbance. At the absorption maximum for DNA i.e. at 260 nm, there is 37% increase in absorption when DNA has completely denatured. This means maximum absorption for ssDNA (single stranded DNA) at A_{260} is 1.37 compared to 1.0 for dsDNA.

8. The temperature at which the DNA molecule is half denatured (melted) i.e. when it has half opened up, is referred to as the melting temperature, (Tm). Tm depends on guanine and cytosine content. The more the guanine and cytosine content, the higher shall be the Tm value. This is because there are three hydrogen bonds between the G:C, the breakage of which requires higher temperature. Therefore, it is the A:T regions of DNA that first melt. Tm also depends on salt concentration. A high salt concentration enhances Tm value. The Tm value ranges from $77^{o}C$-$100^{o}C$ depending upon G:C content which varies from 20%-78% in DNA of various species. On an average Tm value of DNA is taken as $78^{o}C$. 1% of formamide, added to DNA solution, decreases the Tm value by $0.5^{o}C$.

9. The denatured DNA can spontaneously reform the hydrogen bonds between bases and rewind. This is referred to as annealing or renaturation. For renaturation it is required to have a high salt concentration usually 0.15 M or higher of NaCl concentration. The optimum temperature for renaturation is generally $20^{o}C$-$25^{o}C$ lower than the Tm temperature. The renaturation depends on precise collision between complementary strands and hence is a slow process requiring several hours for DNA to fully renature.

10. If multiple DNA molecules are present and denatured by heat, then many single stranded (ss)DNAs will result. Now if the temperature is lowered, the strands will join at random. This molecular mixing is referred to as hybridization; which is a useful method for classification of living organisms based on DNA homology.

11. DNA gives bluish violet color when treated with fuchsin in sulphuric acid.

12. DNA molecules are very long, having thousands of base pairs. Sizes of DNA of few of the organisms are shown in Table 2.3.

13. The molecular weight ranges from 6-100 million.

14. DNA has a density of 1.7 g/ml.

15. The solution of DNA is dextrorotatory.

16. DNA may be linear, circular or super coiled in different biological species as shown in Figure 2.9. Circular DNA means that the chain is continuous end to end. Super coiling means that the double helix (already the two strands are twisted together) further twists over itself to form a super helix. This is important so that DNA assumes a compost shape and is packed into a cell.

17. Supercoiled DNA moves faster than relaxed DNA when centrifuged or electrophoresed.

Table 2.3 : Sizes of DNA of some organisms.

Organism	Base Pairs (in thousand)	Length (µm)	Remarks to note
Polyoma or S.V 40 virus	5.1	1.7	smallest size
E. coli	4,000	1,360	Size of a typical bacteria is 0.3-5 µm
Yeast	13,900	4,600	–
Human	3,184,00	990,000	0.99m; largest in size

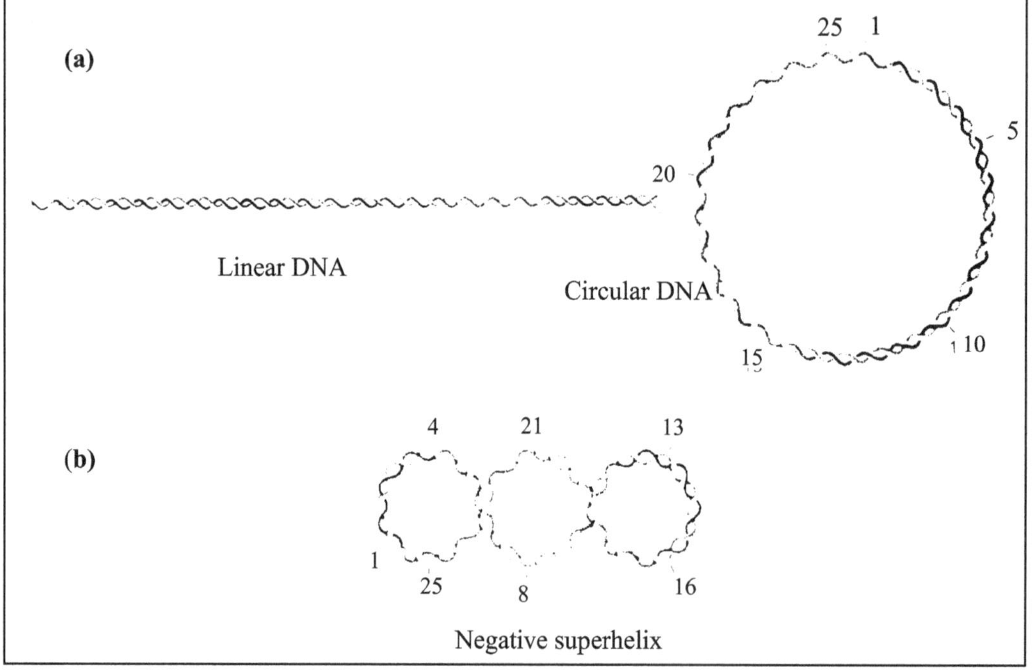

Figure. 2.9 (a) Relaxed and (b) supercoiled forms DNA.

2.6   Replication of DNA

Much of the information concerning the replication of DNA is based on the studies performed in prokaryotes. At the molecular level, the mechanisms of replication and the enzymes involved in replication are the same, both for prokaryotes and eukaryotes. However, some features are unique to the eukaryotes.

2.6.1 Semi conservative Replication of DNA

If you consider the mechanism of replication of DNA at the molecular level, you will see three modes of replication are possible:

1. *Conservative replication :* This means an exact copy (duplication) of the parental dsDNA is expected to remain intact and somehow, the 'progeny' is synthesized.
2. *Semi conservative replication :* This means the hydrogen bonds, holding the two strands together, break resulting in the separation of the two strands. Now both the separated strands act as the parental strand and direct the synthesis based on complementarity of bases i.e. T for A and A for T. The newly synthesized strand is called the complementary strand. Thus in the replicated dsDNA, one strand is parental strand and the other is newly synthesized complementary strand. This refers to the semi conservative mode of replication.
3. *Dispersive replication :* This means there is fragmentation, synthesis and rejoining of the parental and progeny strands. The segments become interspersed.

2.6.1.1   The Meselson and Stahl Experiment

Watson and Crick, based on the structure of DNA, had proposed that the replication of DNA is semi conservative due to base pairing specificity.

This proposal was proved to be true by the experiments of M.S.Meselson and F.W.Stahl in 1958. They grew *Escherichia coli* for several generations in a medium containing the heavy isotope ^{15}N in place of the normal ^{14}N. A DNA which has nitrogenous bases containing ^{15}N has a density of 1.724g/ml, whiles a normal DNA i.e. nitrogenous bases containing ^{14}N, has a density of 1.710g/ml.

After several generations, they transferred the *Escherichia coli* cells, which now contained the 'heavy' ^{15}N DNA, to a medium containing ^{14}N source (Figure 2.10). The cells were allowed to grow for varying periods of time i.e. one generation, two generations etc. and DNA was extracted from the cells after each generation and analysis was carried out in cesium chloride density gradients. After the first generation, the density of DNA was found to be intermediate between the heavy and light type of DNA. After second generation, there was DNA of light and intermediate densities. These results prove the semi conservative replication of DNA.

2.6.2 Enzymes Involved

It was Arthur Kornburg and his colleagues who for the first time in the year 1958 isolated an enzyme from *Escherichia coli* that is responsible for the synthesis of DNA. The enzyme was obviously named as DNA polymerase. As of today there are at least twenty different proteins which take part in the replication of DNA.

2.6.2.1   DNA Polymerases

DNA polymerase that was originally isolated from *Escherichia coli* by Arthur Kornburg is now known as DNA polymerase I, abbreviated as DNA pol I. This is because, a number of enzymes having polymerase activity have been discovered since 1958. These later discovered polymerases are thus named DNA polymerases II and III abbreviated as DNA pol II and DNA pol III. The DNA pol I, II, and III are found in prokaryotes, while in eukaryotes it is DNA pol α, β and γ, that are found and which play corresponding roles in DNA replication.

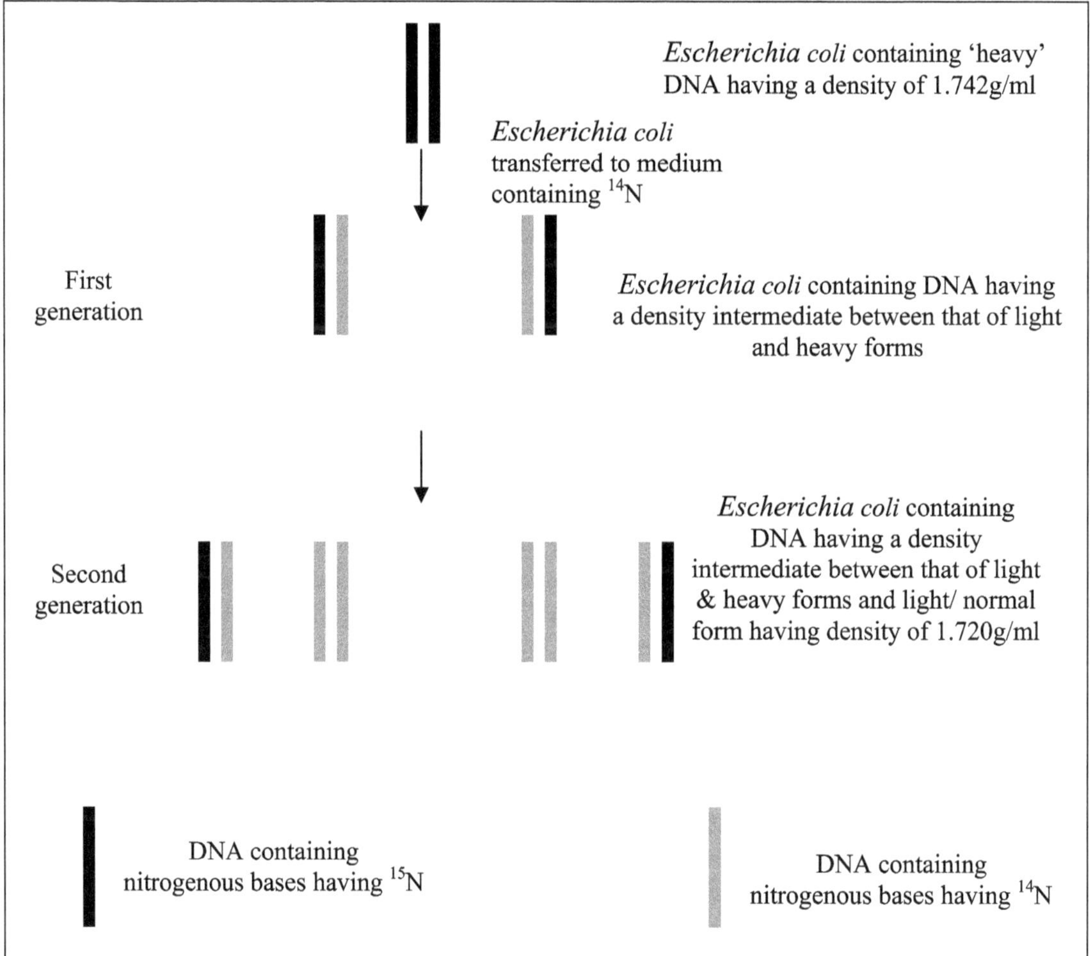

Figure. 2.10 : The Meselson and Stahl experiment : establishing semi-conservative replication of DNA.

Properties of DNA Polymerases

1.   *5′→3′ polymerase activity:* Polymerases bring about the step by step addition of deoxy ribonucleotides at the 3′ end of an existing oligonucleotide.

2. *Nucleophillic attack:* The chain elongation reaction catalyzed by DNA polymerases is a nucleophillic attack of the 3′OH group of the existing oligonucleotide. The reaction involves the formation of a phosphodiester bridge with concomitant release of pyrophosphate.

3. *Obligatory requirement for primer:* Polymerase adds nucleotides on to a pre-existing chain and extends it further. In a way it actually is a transferase enzyme. In fact E.C. (Enzyme Commission) name of the DNA pol I is DNA nucleotidyl transferase. The pre-existing oligonucleotide is referred to as primer. Either a DNA or RNA can act as the primer.

4. *Template-directed:* For its activity i.e. condensation of nucleotide, DNA polymerase has an absolute requirement for ssDNA. It is this ssDNA that will guide as to which nucleotides are to be added. The ssDNA is referred to as the template and the enzyme is referred to as DNA-dependent DNA polymerase.

5. *Requirement of activated precursor:* For the synthesis of DNA, all the four activated precursors must be present. These include deoxyribonucleoside 5′ triphosphates: dATP, dGTP, dTTP, and dCTP.

6. *Requirement for Mg^{2+} ion:* The polymerase also has a requirement for magnesium ion for its activity.

7. *5′→3′ exonuclease activity:* This activity is observed with DNA pol I only and not with DNA pol II or III. This is responsible for catalyzing the removal of primer from the 5′ end of the newly synthesized chain. This activity also is responsible for removal of DNA segments that have been damaged due to ultra violet light irradiation and exposure to other agents.

8. *3′→5′ exonuclease activity or proof reading activity:* This involves removal of newly added nucleotides from the 3′ end of the elongating DNA chain. This activity is opposite to the polymerase (condensation) activity. This activity is responsible for removal of mismatched nucleotides and thus maintains the fidelity of DNA replication i.e. the accuracy of replication without any mistakes in synthesizing the complementary strand.

9. It is to be noted that both the polymerase and exonuclease activities are present in same enzyme. It is only the relative rate of these activities that varies between the three enzymes-DNA pol I, II, and III. Thus DNA pol III has predominant polymerase activity, while DNA pol I has predominant exonuclease activity, while 5′→3′ exonuclease activity is exclusively present in DNA pol I.

10. *DNA pol I (mol.wt.113kDa):* It consists of two bioactive molecules. Its larger fragment (mol.wt. 67 *kDa*) is also known as Klenow polymerase (named after the scientist who discovered it) and possesses both the polymerase and 3′→5′ exonuclease (proof reading) activities, while the smaller fragment (mol.wt. 37 *kDa*) possesses 5′→3′ exonuclease (removal of RNA primer) activity only.

11. *DNA pol II (mol.wt.120 kDa):* Its function is not well understood, however, it is believed to be responsible for DNA repair in prokaryotes.

12. *DNA pol III (mol. wt. >250 kDa):* It is a multimeric enzyme consisting of more than ten subunits which is mainly responsible for DNA replication in prokaryotes. It adds 250-1000 nucleotides per second to the elongating DNA chain.

13. *DNA polymerases in eukaryotes:* In eukaryotes DNA pol α is mainly responsible along with DNA pol δ for DNA replication (predominant polymerase activity) in the nucleus. DNA polymerase ß, assisted with DNA pol ε, is responsible for DNA repair. DNA pol γ is responsible for DNA replication in the mitochondria.

2.6.2.2   Topoisomerases

The word topoisomerase arises from the word 'topology' which is a branch of mathematics dealing with structural properties that are unchanged by deformations such as bending, twisting or stretching.

A key topological property of circular DNA molecule is its linking number, Lk, which is equal to the number of times a strand of DNA winds in the right handed direction around the helix axis. The value of Lk is a sum of twist (Tw) and writhe (Wr). Twist is the number of helical windings of DNA strands around each other, while writhe is number of coilings of the double helical axis over itself. If writhing (super coiling) is right handed than the writhe number carries a negative sign which results in lowering of Lk number. This is referred to as negative super coiling and causes unwinding of the Watson Crick duplex, making it easy for the two strands to separate for replication, transcription or recombination. If writhing (super coiling) is in the left handed direction than writhe number carries a positive sign. This means increase in the Lk number. This is referred to as positive super coiling. This makes the separation of strands much more difficult. Thus in nature DNA super helices are negatively super coiled.

Topological isomers or topoisomers are molecules having different Lk numbers. Thus the relaxed circular form of DNA and the super coiled (either positive or negative) forms of DNA are topoisomers of each other assuming different geometrical shapes. These can be inter converted by cutting (nicking) either one or both the strands of DNA and then rejoining (sealing) of the strands. Enzymes thus catalyzing the inter conversion of topoisomers are called topoisomerases. These were discovered by James Wang and Martin Gellert. Topoisomerases that cleave just one strand of DNA are called Type I topoisomerase, whereas Type II topoisomerases cleave both the strands of DNA.

The Type I topoisomerases catalyze the relaxation of negatively super coiled DNA. The type II topoisomerases catalyze the super coiling of DNA. The type II topoisomerase responsible for super coiling is DNA gyrase. Thus type I topoisomerase and DNA gyrase (a type II topoisomerase) have opposite actions.

Topoisomerases alter the linking number of DNA by catalyzing a three step process:

(i)   Cleavage of one or both strands of DNA,

(ii)   Passage of a segment of DNA through this break,

(iii) Resealing of the DNA break.

2.6.2.3   Helicase

Helicase is also known as unwinding protein, since it is responsible for melting of DNA thereby forming ssDNA. It requires ATP for its unwinding activity.

2.6.2.4   DNA Ligase

DNA ligase joins two fragments of DNA via a phosphodiester bond. For the bond formation, it is required that the two strands should have a free 3′ OH and 5′ PO_4 groups respectively. Note that DNA ligase only closes a nick (cut between two fragments) and cannot incorporate a new nucleotide like DNA polymerase. It is responsible for joining short polynucleotide fragments a (Okazaki fragments) in the lagging strand during DNA replication.

2.6.2.5   Primase

It was mentioned that DNA polymerase III is primer dependent. It can add nucleotides on an existing primer when DNA replication is initiated. Primer is a sequence specific RNA molecule which serves to initiate DNA synthesis. Primase is that enzyme which is responsible for the synthesis of primer.

2.6.2.6   Other DNA binding factors

A number of proteins, about twenty or so, bind with DNA so that replication of DNA could be achieved. These binding factors shall be mentioned as we consider the mechanism of replication.

2.6.3   Modes of DNA replication

It is known that man and eukaryotes in general have more than one chromosome and that the chromosomes are paired E.g. Man has 23 pairs of chromosomes. All bacteria (prokaryotes) on the other hand, have only unpaired chromosome. Further, the DNA in man and eukaryotes in general, is linear while bacterial or prokaryotic DNA in general is circular and super coiled. Circular DNA is also typically present in organelles (mitochondria) of eukaryotes. Viruses may consist of either a linear or circular DNA. Thus based on the geometry of DNA, there are three modes of DNA replication: namely theta, sigma and linear mode.

2.6.3.1   Theta (θ) mode

This takes place in circular DNA. Replication initiates at a specific site called the 'origin of replication', which is specific for each bacterial species. Replication is bidirectional. A bubble forms and increases in size as replication proceeds in both the directions as shown in Figure 2.11a.

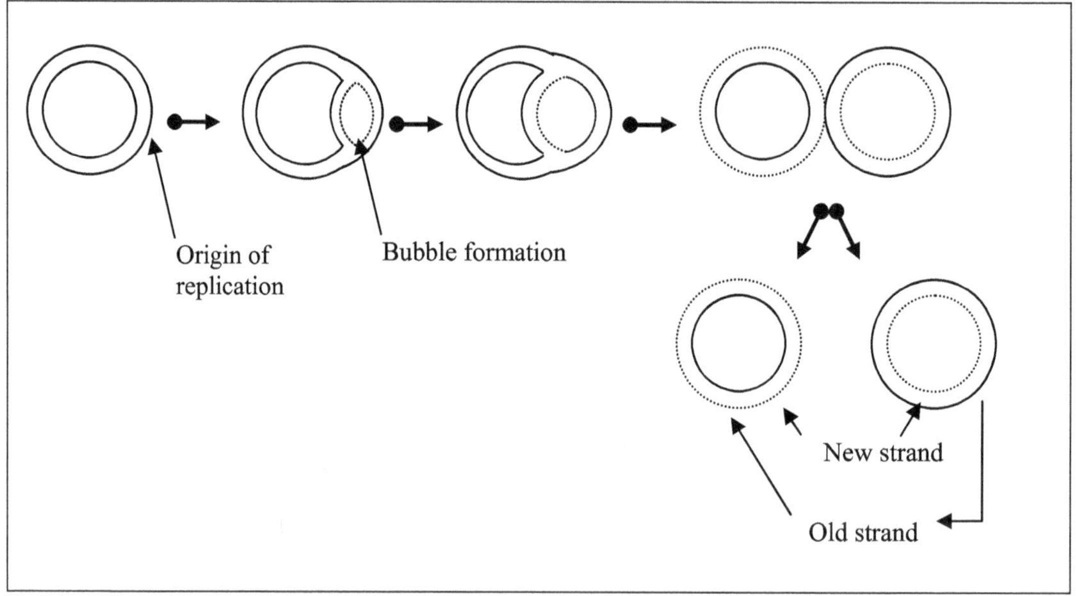

Figure. 2.11 a : Theta mode of replication.

2.6.3.2   Sigma (σ) mode

This also is seen with circular DNA and is also known as rolling circle mode. In this case, one of the strands of dsDNA is nicked (cleavage of phosphodiester bond) thereby exposing $3'OH$ and $5'PO_4$ groups as shown in Figure 2.11b. The circular strand serves as the template. DNA polymerase adds nucleotides at the exposed $3'$ OH terminal, the direction of synthesis being $5' \rightarrow 3'$. The $5'$ PO_4 end of the nicked strand forms the 'tail' of the circle. Replication proceeds further onto this tail. After a short period, an endonuclease nicks the new strand that is complementary to the 'tail' position. The helix further unwinds and replication continues with addition of nucleotides at the $3'$ OH end on circular strand and finally continuing on to add the nucleotides on to the complementary 'tail' position. Again an endonuclease nicks the newly synthesized, short 'tail' position. A ligase joins the two 'tail' fragments. Further unwinding takes place and replication thus proceeds.

2.6.3.3   Linear mode

As the name suggests, this mode of replication is seen with linear DNA. Found in eukaryotes and some viruses. Here, again there is a specific site of origin where the replication initiates. However, there may be hundreds of sites of origin in eukaryotes. The replication may be bidirectional. As replication biotion is initiated, it is seen as a bubble growing in size as replication proceeds. Adjacent bubbles fuse to form larger bubbles. Finally two linear molecules are formed. This mode of replication is depicted in Figure 2.11c.

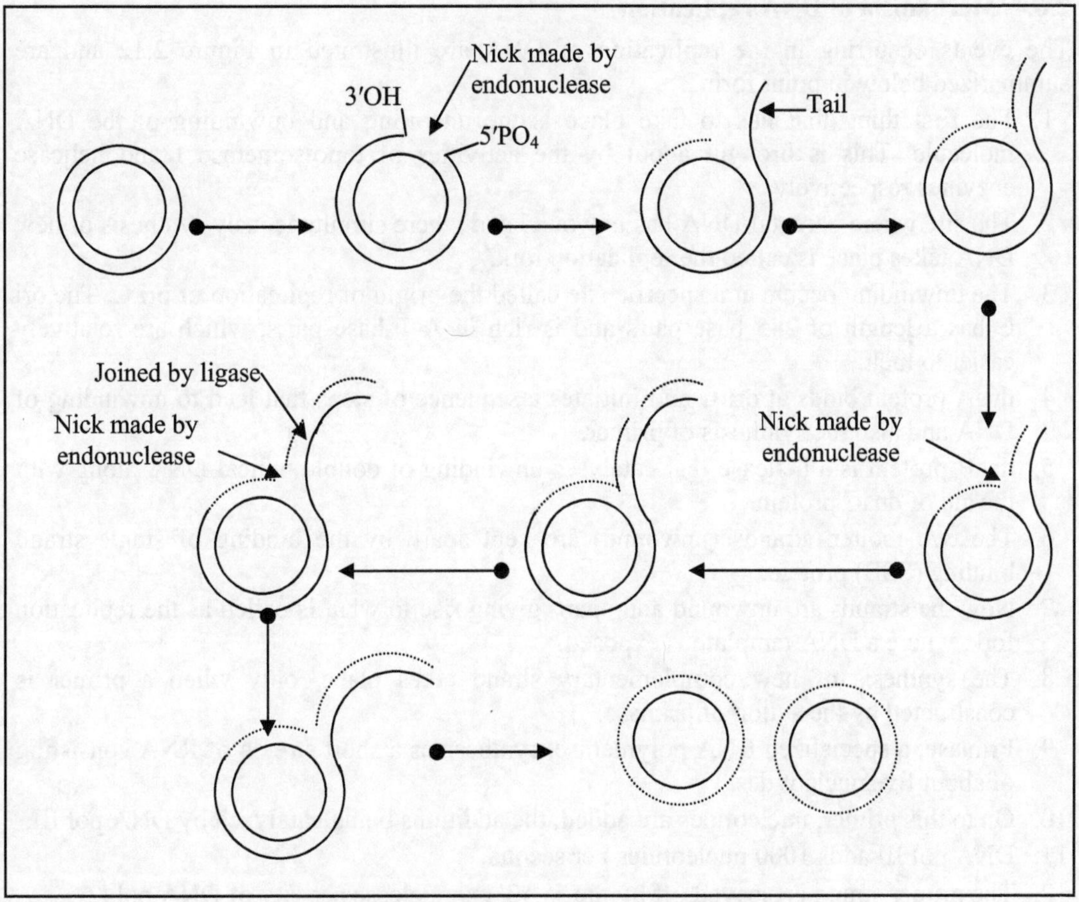

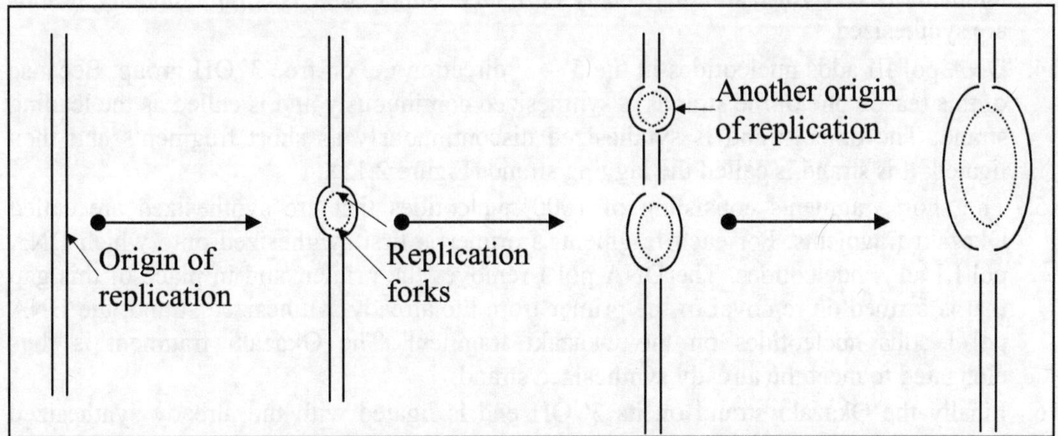

Figure. 2.11 b : Sigma mode of replication.

Figure. 2.11 c : Linear mode of replication.

2.6.4 Mechanism of DNA replication

The events occurring in the replication of DNA are illustrated in Figure 2.12 and are summarized below in point form.

1. The first thing that has to take place is the uncoiling and unwinding of the DNA molecule. This is brought about by the activities of topoisomerase I and helicase enzymes respectively.
2. The site where parental DNA has unwound and where simultaneously synthesis of new DNA takes place is called the replication fork.
3. The unwinding occurs at a specific site called the origin of replication or ori C. The ori C has a length of 245 base pairs and is rich in A-T base pairs, which are relatively easier to melt.
4. dnaA protein binds at ori C and initiates a sequence of steps that lead to unwinding of DNA and also the synthesis of primer.
5. dnaB protein is a helicase that catalyses unwinding of double helical DNA along with the aid of dnaC protein.
6. The two molten strands (unwound) are kept apart, by the binding of single strand binding (SSB) protein.
7. Now the strands are unwound and apart, giving rise to what is called as the replication fork, where a DNA template is exposed.
8. The synthesis of new complementary strand takes place, only when a primer is constructed by the action of primase.
9. Primase, a specialized RNA polymerase, synthesizes a short stretch of RNA consisting of about five nucleotides.
10. On to this primer, nucleotides are added, the additions being catalyzed by DNA pol III.
11. DNA pol III adds 1000 nucleotides per second.
12. The primer strand is cleaved off by the $5' \rightarrow 3'$ exonuclease activity of DNA pol I.
13. The double helical strand ahead of the replication fork keeps on getting unwound by the action of helicase and the two strands kept apart by SSB protein, to serve as template. DNA gyrase simultaneously introduces super coils as complementary strands are synthesized.
14. DNA pol III adds nucleotides in the $5' \rightarrow 3'$ direction i.e. on free 3' OH group. Because of this reason one of the strands is synthesized continuously and is called as the leading strand. The other strand is synthesized discontinuously as short fragments and then ligated; this strand is called the lagging strand (Figure 2.12).
15. The short fragments consisting of 1000 nucleotides that are synthesized are called Okazaki fragments. For each fragment, a primer is first synthesized onto which DNA pol III adds nucleotides. The DNA pol I removes the primer, and in place of this gap that is formed on removal of the primer from the already synthesized strand, the DNA pol I adds nucleotides on the Okazaki fragment. The Okazaki fragment is thus elongated to meet the already synthesized strand.
16. Finally the Okazaki strand on its 3' OH end is ligated with the already synthesized strand on its 5' PO_4 end by the action of ligase.
17. A new loop is then formed, another Okazaki fragment is synthesized, DNA pol I removes the primer and fills the gap, ligase joins the fragment to the strand.

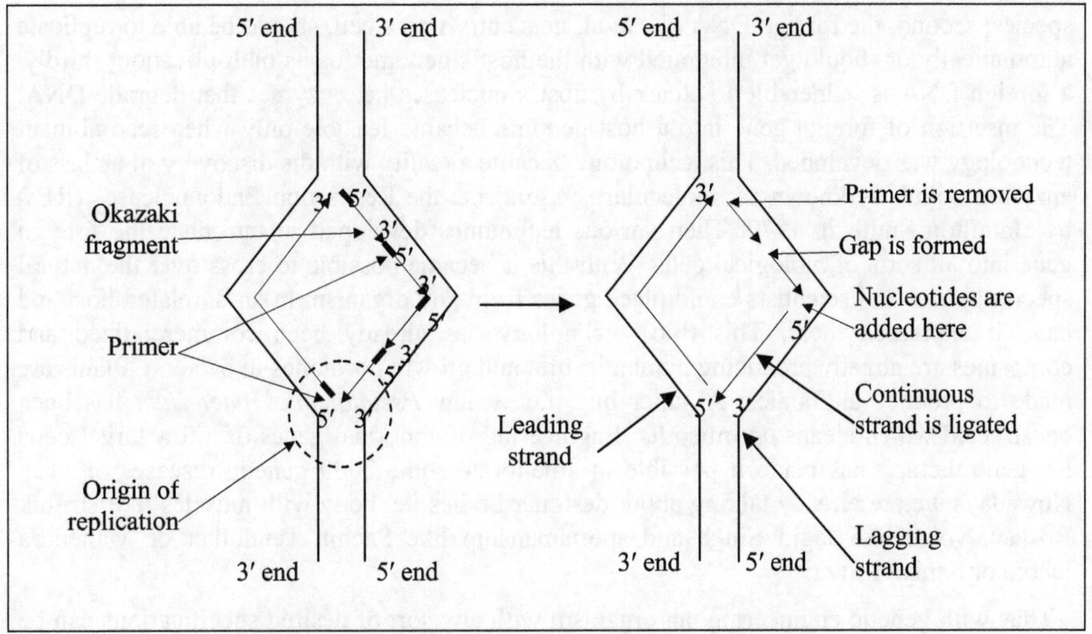

Figure. 2.12 : Mechanism of replication of DNA.

2.7   Engineering the Genes

Sit back for a while, close your eyes and recollect the historical events in the field of genetics. Genes were first detected and analyzed by Gregor Johann Mendel who elucidated the nature of inheritance of characteristics from parents to progeny in living organisms. Then, Frederick Griffith and O.T. Avery, C.M. Macleod & M.Mc Carty proved that DNA is the hereditary material based on their experiments that some bacterial cells could spontaneously (by transformation) take up fragments of DNA. J.D Watson and F.H.C. Crick deduced the structure of DNA. At the molecular level, genes are arranged in a linear fashion on the long DNA molecule. The genetic code was deciphered by Nirenberg & Holley and Khurana. It was found that a code for amino acids consists of a triplet codon, there being a total of twenty amino acids and 4^3=64 codons (4 nitrogenous bases arranged in triplet). Therefore more than one codon could code for a given amino acid; this is known as degeneracy in genetic code.

The most important point to note here is that the genetic language is universal. It is made of only four letters A, G, C and T and only 64, three letter words i.e. the 64 codons, encoding for twenty amino acids making up all the vast array of proteins in all the organisms.

Thus, having understood the Natural Laws of Genetics, scientists tried to engineer genes so as to derive useful products (i.e. biotechnology). To do so, cell free 'naked' DNA was added to a culture of microbial, plant and animal cells, with the intention that transformation could take place. But success was not achieved. Why? Because of three reasons: first, transformation spontaneously occurs only in bacteria that too in some and not all the species; second, the foreign DNA if it at all gets entry into a cell, should be able to replicate automatically or should get integrated with the host's genome for its multiplication; thirdly, a foreign DNA is vulnerable to attack by host's nucleases i.e. enzymes that degrade DNA. The insertion of foreign gene into a host genome became feasible only when recombinant technology was developed. This technology became a reality with the discovery of a class of enzymes popularly known as 'molecular scissors' i.e. the Restriction Endonucleases (REs) by Hamilton Smith in 1970. Then various techniques developed to introduce the foreign gene into all sorts of biological cells. With this it became possible to cross over the natural species barrier and scientists could place genes from any organism in an unrelated host and have it expressed there. This (bio) technology has already been commercialized and companies are already producing human insulin and growth hormones in bacteria. Plants are made to produce antibodies. A super bug, i.e. a new *Pseudomonas bacterium* has been constructed which cleans up oil spills. Engineering of abnormal genes of a few target cells i.e. gene therapy has made it possible to ameliorate some 2000 genetic diseases of man. Nowadays we are already talking about designer bodies i.e. body with muscles like Hrithik Roshan, voice like Jagjit Singh and sportsmanship like Sachin Tendulkar or Mahendra Dhoni or Saniya Mirza!

Thus with genetic engineering, an organism with any sort of desired specifications can be produced. Only future will tell as to where this technology will lead us to.

This technology has turned into reality only because of the necessary enzymes present in nature which have been discovered and put to judicious use by man. We shall now turn our attention to the enzymes involved in genetic engineering.

2.7.1 Enzymes involved

As already introduced in the preceding section, it is now possible to identify and isolate the desired gene, coding for the protein of interest, and using a suitable vehicle, insert it into the genome of another unrelated organism. But this is a complex, multi step process involving calculated manipulation of the donor, vector and host genomes. Some of the commonly used enzymes involved in the genetic engineering manipulation are considered below.

2.7.1.1 Restriction Endonucleases (REs)

Genetic engineering became a possibility only after the discovery of restriction enzymes. REs cut dsDNA molecules internally (endonuclease) at specific (restriction) recognition sites. The first RE was discovered by Hamilton Smith in 1970, in the bacterium *Haemophilus influenzae*. Since then several hundred REs have been discovered.

REs are found in bacteria where they serve to cut (digest) the invading foreign DNA. They recognize a specific sequence within the DNA molecule. The sequence may consist of 4, 5, 6, 7 or 8 nucleotides in length. They then cut the DNA backbone at the site within the DNA molecule. Hence they are also known as 'molecular scissors'. REs are of three types: Type I, Type II and Type III. Type I RE recognizes a specific site but the cleavage site is some 1000 nucleotides away from the recognition site. Hence they are not preferred to be used in genetic engineering. In genetic engineering, it is the type II REs that are employed. The type II REs recognize and cleave at the same specific site or within the same specific sequence.

Mg^{2+} is required for their activity. Mn^{2+} can serve as a replacement for Mg^{2+}. Besides Mg^{2+}, an optimum salt concentration is also required that may be a high concentration of 150 mM NaCl, medium concentration of 50 mM NaCl or a low concentration of <10 mM NaCl for different restriction endonucleases'. Restriction endonuclease from SmaI is an exception in that it requires KCl instead of NaCl for its activity. Further if the site is methylated, it will not be recognized and will not be cleaved by the enzyme. It is to be noted that one particular restriction endonuclease recognizes only one specific sequence, however the sequence is not unique (reserved) for that particular restriction endonuclease. The same sequence can be recognized by more than one enzyme. Such enzymes recognizing the same specific sequence but digest at different place are called isoschizomers.

A typical example of type II restriction endonuclease is Eco RI. The REs are named depending upon their bacterial origin in the following manner:

- First letter 'E' stands for the genius.
- Next two letters 'co' stand for the species.
- The fourth letter 'R' stands for the strain.
- The Roman number 'I' stands for the numerical order of isolation if more than one restriction endonuclease has been isolated from particular bacteria. This means there is Eco RII, i.e. the second restriction endonuclease isolated from *Escherichia coli* of R strain.

There are two possible ways as to how restriction endonucleases may cut the dsDNA molecule at specific recognition site. The cut can be made symmetrically or asymmetrically through the site as shown in Figure 2.13.

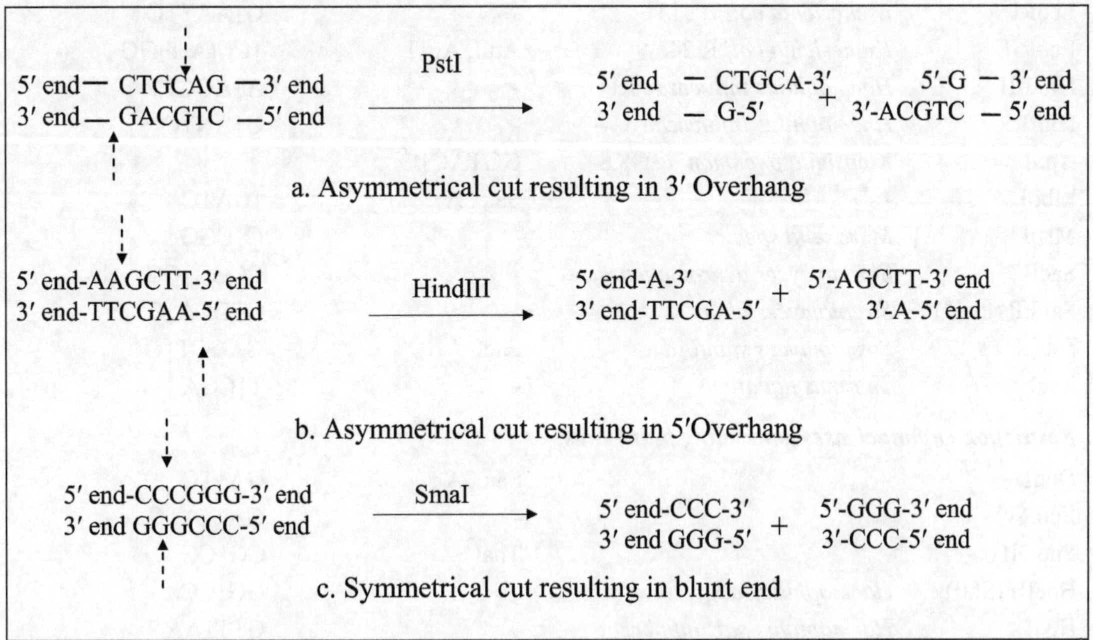

Figure. 2.13 : Asymmetrical and symmetrical cuts made by restriction endonucleases.

When the cut made in the backbone is symmetric, it results in the formation of fragments that are said to have square or blunt ends. When the cut made in the backbone is asymmetric, it results in the formation of fragments that are said to have sticky or cohesive ends. If the cut is asymmetric, it means that the site of digestion may be either towards 5′ end or 3′ end of the recognition sequence, thereby generating a small region of over hanging ssDNA referred to as 5′ overhang or 3′ overhang respectively. From the point of view of genetic engineering, REs that generate sticky ends are useful.

A brief listing of the REs is given in Table 2.3. It will be interesting to note that all the recognition sequences are palindromes, i.e. to say, the complementary nucleotide sequences are same as the original sequence when read in the reverse direction. In Table 2.3, the single strand sequences are given; do a little exercise for your self and write out the complementary sequence as it would occur in the template strand. You will see that the sequences in the two strands are identical when read in the forward and backward direction i.e palindromic.

Table 2.3 : A brief listing of restriction endonucleases.

Enzyme	Microbial Source	Isoschizomers	Recognition sequence
Restriction endonucleases generating cohesive ends			
AatII	*Acetobacter aceti*	-	GACGT↓C
ApyI	-	AtuI, EcoRII	CC↓(A/T)GG
AsuI	-	-	G↓GNCC**
BclI	*Bacillus caldolyticus*	-	T↓GATCA
EcoRI	*Escherichia coli* RY13	-	G↓AATTC
EcoRII	*Escherichia coli* R 245	AtuI, ApyI	↓CC(A/T)GG
HindIII	*Haemophilus influenzae* RD	-	A↓AGCTT
HinfI	*Haemophilus influenzae*	-	G↓ANTC**
KpnI	*Klebsiella pneumoniae* OK8 -	GGTAC↓C	
MboI	-	Sau 3A	↓GATC
MspI	*Moraxella species*	-	C↓CGG
SacII	*Streptomyces achromogenes*	-	CCGC↓GG
Sal PP (PstI*)	*Streptomyces albus*	-	CTGCA↓G
SstI	*Streptomyces stanford*	SacI	GAGCT↓C
TaqI	*Thermus aquaticus*	-	T↓CGA
Restriction endonucleases generating blunt ends			
DpnI	-	Sau 3A	GA↓TC
EcoRV	-	-	GAT↓ATC
FnuDII	-	ThaI	CG↓CG
HaeIII (SfaI)	*Haemophillus aegyptius*	-	GG↓CC
HpaI	*Haemophilus parainfluenzae*	-	GTT↓AAC
SmaI	*Serratia marcescens* Sb	XmaI	CCC↓GGG

**Pst: Providencia stuartii; **N: no base prefernece*

2.7.1.2   DNA Polymerases I

The detailed properties of this enzyme have been discussed earlier. In genetic engineering it is used because of its $5'{\rightarrow}3'$ and $3'{\rightarrow}5'$ exonuclease activities for the purpose of synthesizing the complementary strand of DNA or cDNA.

2.7.1.3   Klenow Polymerase

This is the large fragment of DNA pol I possessing $5'{\rightarrow}3'$ polymerase and $3'{\rightarrow}5'$ exonuclease activities. This is also employed for synthesizing cDNA.

2.7.1.4   T₄ DNA Polymerase

This polymerase has $5'{\rightarrow}3'$ polymerase and $3'{\rightarrow}5'$ exonuclease activities like Klenow polymerase. T_4 DNA polymerase has a marked $3'{\rightarrow}5'$ exonuclease activity in comparison to the above polymerase. Also its $3'{\rightarrow}5'$ exonuclease activity is significant on ssDNA. Therefore, it is employed for digesting the $3'$ overhang formed by RE.

2.7.1.5   Reverse Transcriptase (RTase)

The source of this enzyme earlier was the Avian myeloblastosis virus (AMV) or from MML virus. However, now the genes encoding for RTase have been cloned and the enzyme is thus obtainable by rDNA technology. The enzyme catalyzes the synthesis of copy DNA (cDNA) based on the genetic information present in ss mRNA. Therefore, it is an RNA dependent DNA polymerase. However, a ssDNA can also serve as its template. Interestingly, the synthesis of ssDNA (which is based on an RNA template) itself involves the use of this enzyme. In the subsequent step this synthesized ssDNA serves as the template for the synthesis of the second strand of cDNA. For initiation of its activity, the enzyme requires a DNA primer (compare with DNA polymerase) complementary to the RNA template and Mg^{2+} or Mn^{2+}. It has an optimum activity at a pH of 8.3.

2.7.1.6   Alkaline Phosphatase

This is a zinc containing enzyme and when it is isolated from *Escherichia coli* it is called as bacterial alkaline phosphatase (BAP) and when isolated from calf intestine it is called as calf intestine phosphatase (CIP). BAP has higher activity in comparison to CIP. The enzyme dephosphorylates i.e. removes $5'$ terminal phosphate from nucleic acids to generate free $5'$ OH group. The dephosphorylation prevents recircularization of vectors, so that they remain linearized. Also, it is at this free $5'$ OH group that DNA can be labeled by phosphorylation with polynucleotide kinase and γ-^{32}P ATP to produce ^{32}P- labeled DNA.

2.7.1.7   DNA ligase

The activity of DNA ligase is just opposite of RE. If RE is called 'molecular scissor' then DNA ligase is called 'molecular suture'. This enzyme catalyses the joining of two different pieces of DNA. The source of the enzyme is either a T_4 virus or *Escherichia coli* bacteria.

The joining of two DNA fragments is brought about by formation of phosphodiester bond between the free 5'-PO$_4$ group on one strand and 3'-OH group on the other strand. The enzyme requires ATP for this energy dependent activity. It is to be noted here, that the cohesive or sticky ends generated by RE (in the desired gene or vector) are more efficiently joined than the blunt or square ends by the ligases. To bring about efficient ligation of the blunt-ended DNA pieces, a high concentration of T$_4$ DNA ligase is required or its activity can be stimulated about twenty times by the addition of T$_4$ RNA ligase. The enzyme is needed to join the desired fragment of DNA (desired gene) with the nicked vector DNA. Note that DNA ligase joins pieces of DNA via phosphodiester bond formation, however, unlike DNA polymerase; it cannot add nucleotides on to a DNA fragment.

2.7.1.8   T$_4$ Polynucleotide Kinase

Again this is an enzyme that has an activity opposite to that of alkaline phosphatase. Polynucleotide kinase catalyses the addition of a phosphate group at the free 5'OH group of a nucleic acid. It is important to note that it is only the γ-PO$_4$ of ATP (and no other nucleotide) that is transferred. This enzyme is thus used to label nucleic acids with ^{32}P.

2.7.1.9   Terminal deoxynucleotidyl transferase

This enzyme catalyses the addition of a chain of 40 or more commonly 20 nucleotides, referred to as 'tail', on to the 3' end of nucleic acids. In genetic engineering, it is used to add a 'tail' to cDNA and the nicked vector DNA so that the two can be recombined by a technique called as homopolymer tailing. It can also be used to label nucleic acids at the 3' end.

2.7.1.10 DNase I

This is a nonspecific endonuclease that cleaves both ssDNA and dsDNA. The enzyme can be used to obtain pure RNA, from a mixture of RNA and DNA. It is also used to create nicks in dsDNA during nick translation.

2.7.1.11 RNase H

This is a specific enzyme that digests RNA. It is used to obtain pure DNA, when it is contaminated with RNA. In genetic engineering, its job is to digest the RNA strand, once the cDNA is synthesized.

2.7.1.12 Methylases

These enzymes, like REs, recognize specific nucleotide sequence, and transfer a methyl (–CH$_3$) group at the recognition site of DNA. A methylated site in the DNA cannot be cleaved by an RE. Thus, methylases are used to protect the internal restriction sites in genetic engineering.

2.7.1.13 S_1 Nuclease

It is a specific, ssDNA endonuclease that cleaves DNA to remove 5′ nucleotides from the 5′ end. Normally the enzyme cannot hydrolyze the double stranded nucleic acid. The enzyme requires Zn^{2+} and a pH = 4.45 for its optimum activity. The enzyme cleaves single stranded regions in duplex DNA such as hairpin loops and gaps and torsional stress points in super helical DNA.

Having taken a brief look at some of the biocatalysts involved in engineering of genes, we can now focus out attention to the actual process of genetic engineering.

2.7.2 Steps of genetic engineering

The job of a gene engineer is to identify and isolate the gene of interest and then have it introduced and joined in the genome of another organism wherein it can either be cloned or allowed to express itself to produce the desired gene product.

The steps involved in genetic engineering can be enumerated as:

1. Isolation of gene of interest, which is referred to as foreign or target gene.

2. Insertion of isolated gene into a vector to form a recombinant DNA (rDNA) molecule.

3. Introduction of rDNA into a host cell.

4. Identification of and isolation of the transformed cells i.e. cells which have taken up the rDNA (foreign gene + vector).

5. Cloning i.e. replication of rDNA to produce multiple copies of itself.

6. Expression of foreign gene to obtain the desired gene product.

7. Now we shall consider the details stepwise.

2.7.2.1   Gene of interest

The gene of interest that has to be isolated may be of bacterial, plant or animal origin or it can even be newly (invented) synthesized gene. There are various ways and means of obtaining the gene of interest. These are discussed below.

1. **Genomic DNA :** In this method, DNA that is enclosed within the cell has got to be released by treating with enzymes, surfactants or by subjecting to pressure. The released DNA is then separated from other cellular components by way of centrifugation.

   Now to obtain the desired gene from the long stretch of DNA, restriction enzymes are used. We have already discussed these enzymes under section 2.7.1.1. It is to be mentioned here that these enzymes recognize specific sequences in long DNA stretch and cleave it. Thus, by using the appropriate RE, i.e. one that recognizes the sequences just before and/or after the desired gene, DNA fragments or genes will be obtained. These are then separated from the not required fragments by way of sucrose-density gradient sedimentation.

2. **Complementary/copy DNA (cDNA)** : Genes in eukaryotic cell (plants, animals) consist of introns and exons. Introns are segments of DNA that do not code for amino acids. It is the exons, i.e. segments of DNA, that code for amino acids. Introns and exons are found alternating in a gene. When a gene has to be expressed, it is transcribed to mRNA (both with introns and exons). However, before mRNA leaves the nucleus to go into the cytoplasm, the introns are removed by the cell machinery, this is called as splicing.

    Unlike the eukaryotic gene, a prokaryotic gene (bacteria) consists only of exons. Thus, gene splicing does not occur in prokaryotes.

    Now if a eukaryotic gene (with introns and exons) is introduced into a prokaryotic host's genome, a functional gene product could not be expressed. This is because splicing does not occur in prokaryotes. This problem is solved by using cDNA.

    In case of using genomic DNA as the source for desired gene, it is essential to isolate DNA then locate its position and have it recognized and cut using REs. But in this case, mRNA, a transcript of DNA, coding for a particular (desired) protein, such as mRNA for haemophilia, serves as the template for reverse transcription to synthesize a copy DNA (see sections 2.7.1.5 & 2.7.1.8). The construction of cDNA is diagrammatically presented in Figure 2.14.

3. **Chemical Synthesis** : The gene of interest can be chemically synthesized if the amino acid sequence in the desired protein is known. Using this technique, it is even possible that entirely a new gene coding for a new protein, that has never been discovered or is not found in the universe, can be synthesized!

4. **Mechanical Shearing** : This involves subjecting of DNA to high speed mixing and sonication so that target fragments of DNA are generated. But this process necessarily involves the use of homopolymer tailing (see section 2.7.1.8).

2.7.2.2   Insertion into vector

A vector is a segment of DNA with which the foreign gene can be ligated using a DNA technique and carries it into a cell. A vector contains the following genetic elements:

- Origin of replication i.e. it has to be self replicating;
- Selectable genes: - which code for a property that allows a cell to grow under hostile conditions e.g. gene coding for antibiotic resistance or on nutrient deficient medium. This allows identification of cells that have been successfully transformed;
- Polylinker: - it is a piece of DNA containing cleavage sites for a number of REs. This allows one of the REs to cleave and insert the desired gene;
- Specialist origin: - this allows the vector (such as plasmid) to have a high copy number; and
- promoter, enhancer and leader peptides: - these regulate expression of foreign gene in the host.

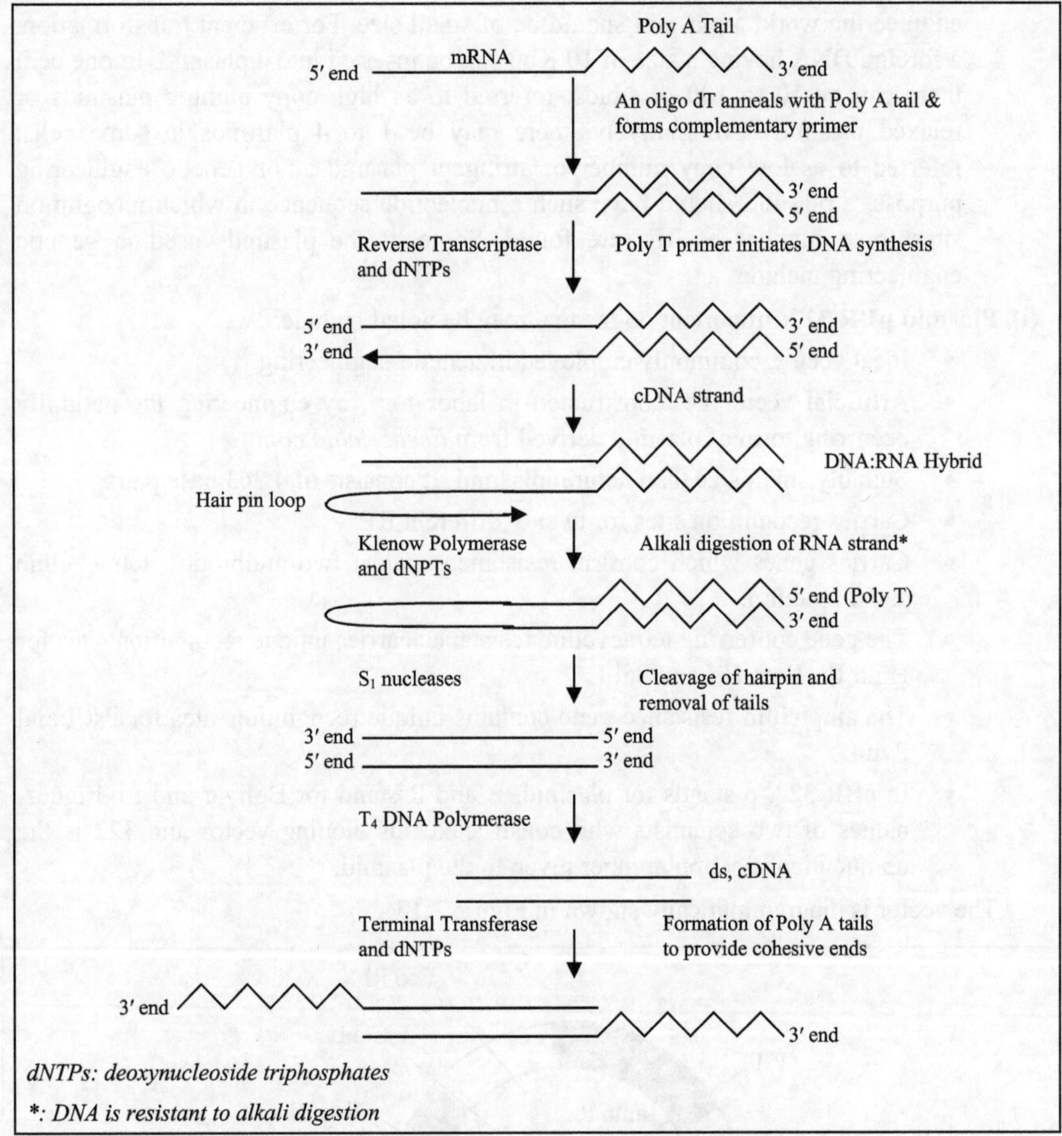

Figure. 2.14 : Construction of double stranded cDNA from mRNA by S_1 nuclease method using Klenow polymerase.

The vectors that are used for genetic engineering work include :

1. **Plasmids :** Plasmids are extra chromosomal, self replicating, double stranded, circular DNA molecules present in bacteria. The plasmids carry genes that are responsible for the formation of bacteriocins, confer resistance against a number of antibiotics or control the synthesis of sex pilli. The size of plasmids ranges from 1 to more than 500 Kbp (1Kbp = 1000 nucleotide base pairs). Ideally, for genetic

engineering work, a plasmid should be of small size. For efficient transformation, a foreign DNA having a size of 10 Kbp can be inserted into a plasmid. In one cell, there can be 10 to 100 plasmids, referred to as high copy number plasmids or relaxed plasmids. Alternatively, there may be 1 to 4 plasmids in some cells, referred to as low copy number or stringent plasmids. For genetic engineering purpose, a plasmid should have such a nucleotide sequence in which recognition sites for a number of REs are found. Some of the plasmids used in genetic engineering include:

(i) Plasmid pBR 322: Important points that may be noted include:

- Ideal vector, commonly employed in genetic engineering.
- Artificial vector i.e. constructed in laboratory, by engineering the naturally occurring form of plasmid derived from *Escherichia coli*.
- Suitably small size than natural plasmid. It consists of 4,363 base pairs.
- Carries recognition sites for twenty different RE.
- Carries genes which confirm resistance against two antibiotics: tetracycline and ampicillin.
- The gene conferring tetracycline resistance carries unique recognition sites for Hind III, Bam H I and Sal I.
- The ampicillin resistance gene contains unique recognition sites for Pst I and Pvu I.
- In pBR 322, p stands for plasmid, B and R stand for Bolivar and Rodriguez, names of two scientists who constructed this cloning vector and 322 is the unique identification number given to the plasmid.

The vector is diagrammatically shown in Figure 2.15

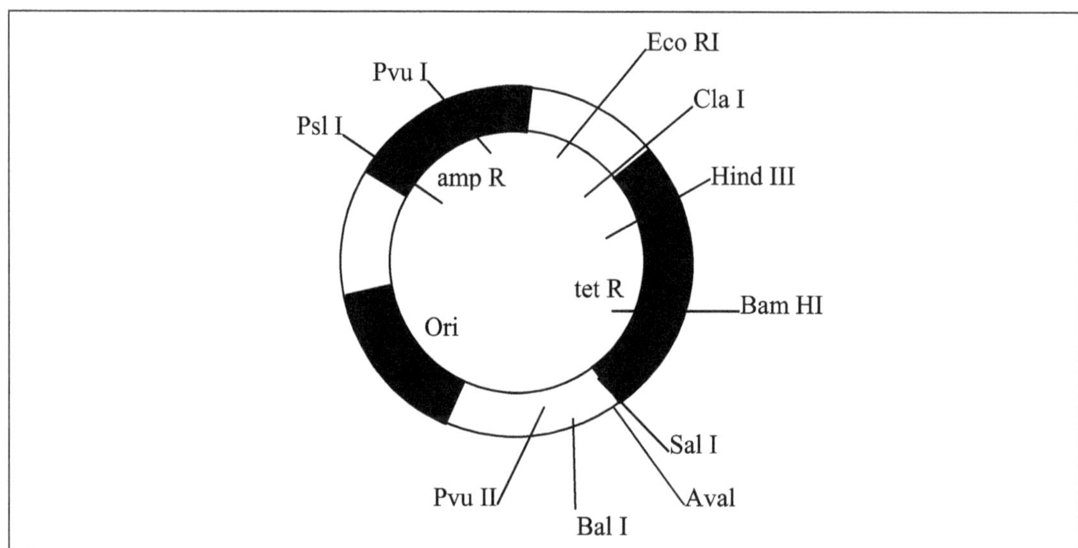

Figure. 2.15 Vector: pBR 322.

(ii) Plasmid pUC 19 : Important points that may be noted include:

- Artificially constructed vector.

- p stands for plasmid, U and C stand for University of California, where this vector was constructed by two scientists named J. Messing and J. Viera and 19 is the unique number given for distinguishing it from other plasmids.

- Contains genes for ampicillin resistance, ß-galactosidase (lac Z) and lac I.

- The lac Z gene contains unique recognition sites for Eco RI, Sac I, Xma I, Sma I, Bma H-I, Pst I, Hind III etc.

- It consists of 2680 base pairs.

- It offers an easier and less time consuming screening process for selection of transformants in comparisons to pBR 322.

- It has a high copy number of up to 500 per cell.

2. **Bacteriophage-** Bacteriophage is a virus that infects bacteria. Unlike plasmids, bacteriophages can accommodate larger pieces of DNA (i.e. large sized desired genes). The bacteriophages that are used as cloning vector include:

(i) Bacteriophage lambda (λ): Important points that may be noted include:

- It infects *Escherichia coli*.

- Foreign genes of 8.9 to 22.4 Kbp in length can be inserted into it.

- The phage DNA is linear, double stranded and 49 Kbp long.

- The bacteriophage λ DNA can be imagined to be divided into left arm, middle and right arm genetic regions. The genes in left arm are responsible for synthesis of head and tail. The genes in right arm are responsible for regulation, immunity, DNA synthesis and host lysis. The middle region is responsible for integration and excision during lysogenic cycle of virus. During lysogenic cycle, the viral DNA after entry into a bacterial cell, remains silent i.e. does not replicate for some period of time. This lysogeny is not desired here in genetic engineering work. Therefore, the middle region, comprising of 40 percent i.e. about 20 Kbp can be removed and replaced with the foreign gene. This will allow viral DNA (with foreign gene) to enter into lytic cycle i.e. where replication occurs. (see Figure 2.16)

- A single stranded, twelve nucleotide long region is present on the 5'ends of both the strands. The protruding (over hanging) regions are complementary to each other. These regions are called cos sites which provide cohesive ends and allow for circularization of the phage DNA.

- After transduction, once the phage DNA is introduced, the linear, phage DNA, circularizes due to annealing of cos sites and replicates by theta/ rolling circle modes.

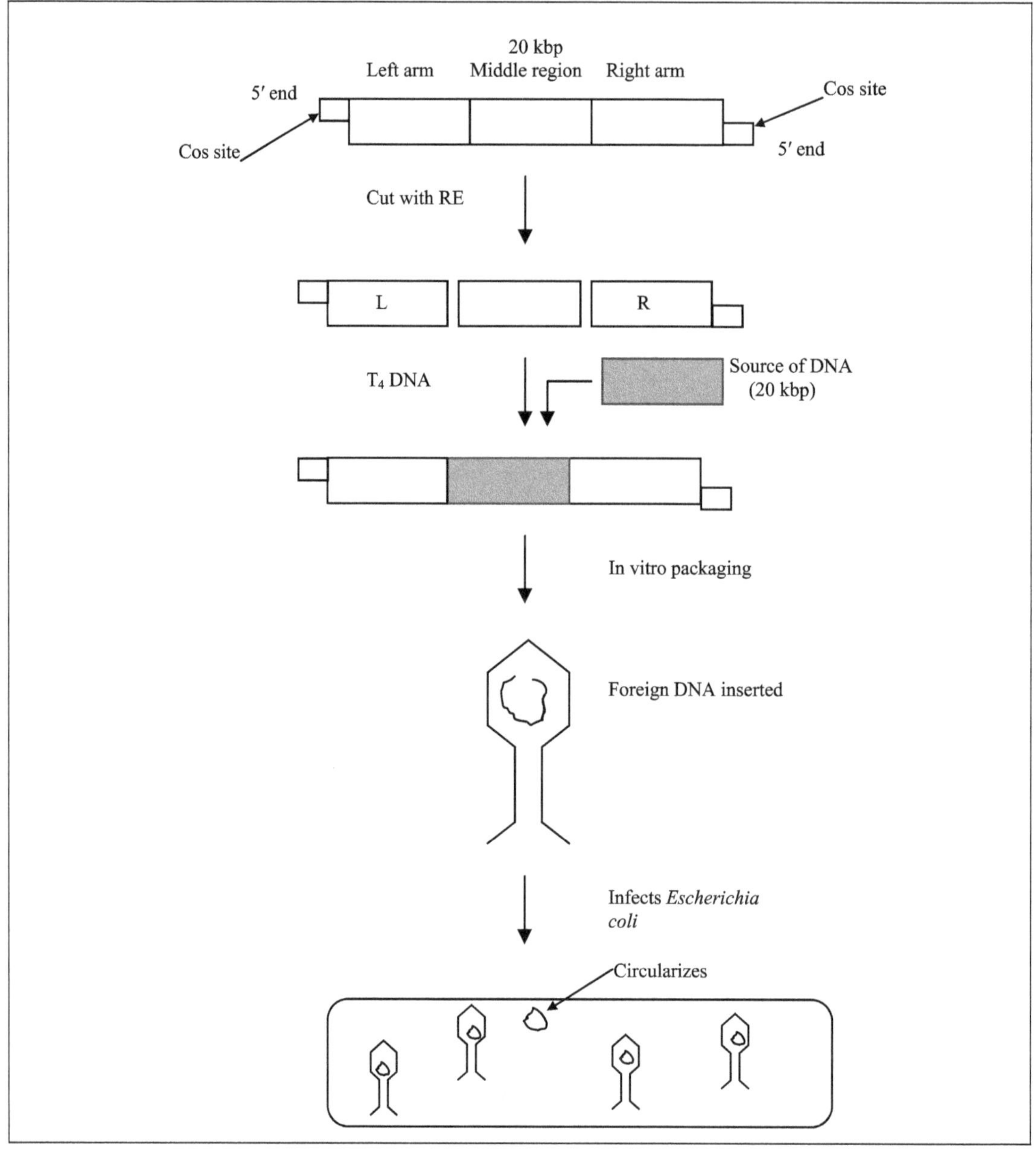

Figure. 2.16 : Lambda phage DNA.

(ii) M 13 Bacteriophage: Important points that may be noted include:
- Phage DNA is single stranded.
- Infects only F^+ cells.
- Comprises of 6402 base pairs.
- A lac Z gene using which transformants can be screened is introduced into a non-coding region of M 13.

3. **Cosmids or phagemid-** A cosmid is a hybrid vector that has been engineered to combine the properties of a bacteriophage λ (its cos sites) and the plasmid (its origin of replication site and tetracycline resistance gene); hence, it is named cosmid, '*cos*' from cos site and '*mid*' from the plasmid (see Figure 2.17). Thus, it is also known as phagemid. It offers six restriction sites. Most importantly it can carry a foreign gene of about 40 Kbp. Like a plasmid, it is self replicating and like a bacteriophage, it infects by transduction.

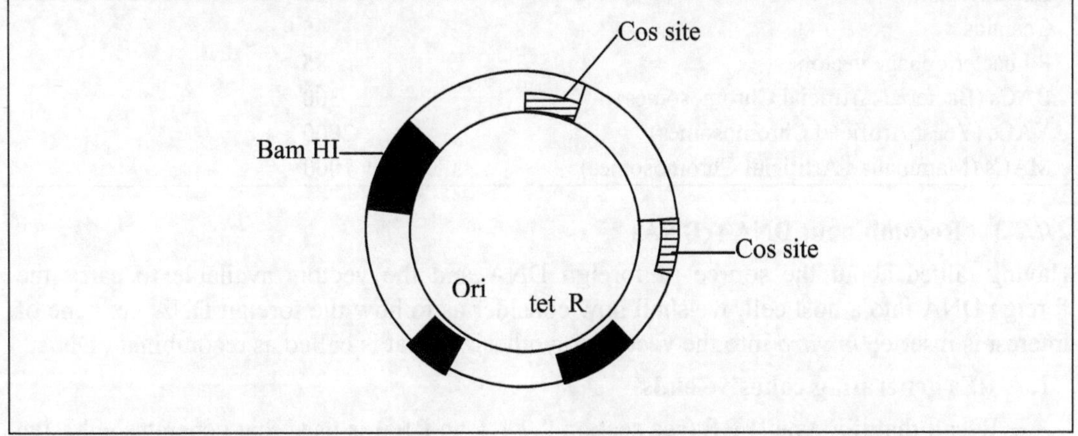

Figure. 2.17 : Diagrammatic representation of a cosmid.

4. **Shuttle vectors-** These replicate in different hosts such as *Escherichia coli* or yeast which is a eukaryotic host. The advantage of using a eukaryotic host is that it has cellular machinery for post translational modification, which is a must when eukaryotic proteins are produced. Examples of shuttle vectors include

    (i)   Yeast episomal plasmid

    (ii) Yeast artificial chromosome

5. **Zoophaginea -** These are viruses that infect animals. In genetic engineering, these virus are used to introduce a foreign gene into animal host cells such as monkey cells or insects. Zoophaginea used as vectors include:

    (i) SV_{40} : Important points to be noted:
        * Simian 40 virus consists of double stranded, circular DNA.
        * It is used to introduce foreign genes into monkey cells.

    (ii) Retrovirus :
        * Introduces foreign DNA into mammalian cells.

    (iii) Baculovirus :
        * Introduces foreign DNA into insect cells.

6. **Phytophaginea –** These are virus that infect plants and are used to introduce foreign DNA into plants for expression. Examples include:

    (i)   Tomato golden mosaic virus

    (ii) Cauliflower mosaic virus

The various vectors and the sizes of foreign DNA that can be packed into them are presented in Table 2.4.

Table 2.4 : Vectors and their maximum hold size for foreign DNA.

Type of vector	Typical maximum "insert" size (in Kbp)
"Conventional plasmid vectors"	15
Lambda phage vectors (including "phagemids")	20
Baculovirus	20
Cosmids	45
P I bacteriophage vectors	85
BACs (Bacterial Artificial Chromosomes)	300
YACs (Yeast Artificial Chromosomes)	2000
MACs (Mammalian Artificial Chromosomes)	At least 1000

2.7.2.3   Recombinant DNA (rDNA)

Having talked about the source of foreign DNA and the vectors available to carry the foreign DNA into a host cell, we shall now consider as to how the foreign DNA i.e. gene of interest is inserted *in vitro* into the vector to synthesize what is called as recombinant DNA.

1. **REs generating cohesive ends**

    Recall that if a type II RE (see section 2.7.1.1 on REs) is used that generates cohesive ends and that if the same RE is used to cut and obtain the desired gene and is also used to make a cut in the vector, then the staggered ends that are generated in the two (desired gene and vector) would have a complementary sequence. If the vector and desired genes are brought together, *in vitro* annealing would occur spontaneously; further the ends of the two DNAs can be joined using DNA ligase (2.7.1.6) at a temperature of 4°C-11°C requiring 12-24 hours. The construction of recombinant molecule using a RE is depicted in Figure 2.18.

2. **RE generating blunt ends**

    Ligation of desired gene and vector having blunt ends can be brought about by using a high concentration of DNA ligase than that required for ligating cohesive ends. It is the T_4 DNA ligase that has to be employed rather than *Escherichia coli* DNA ligase to join blunt ends.

3. **Homopolymer tailing**

    By using a RE, if blunt ends are generated then the homopolymer tailing technique is useful for inserting the desired gene into vector. This technique involves the use of terminal deoxynucleotidyl transferase (TdT) (see section 2.7.1.8). The concept is depicted schematically in Figure 2.19.

    The technique involves the use of an RE, say Hpa I, to cleave a circular plasmid at its specific cleavage site to generate a linear plasmid i.e. vector. Likewise, a restriction enzyme fragment or cDNA i.e. desired gene, is obtained which will also have blunt ends.

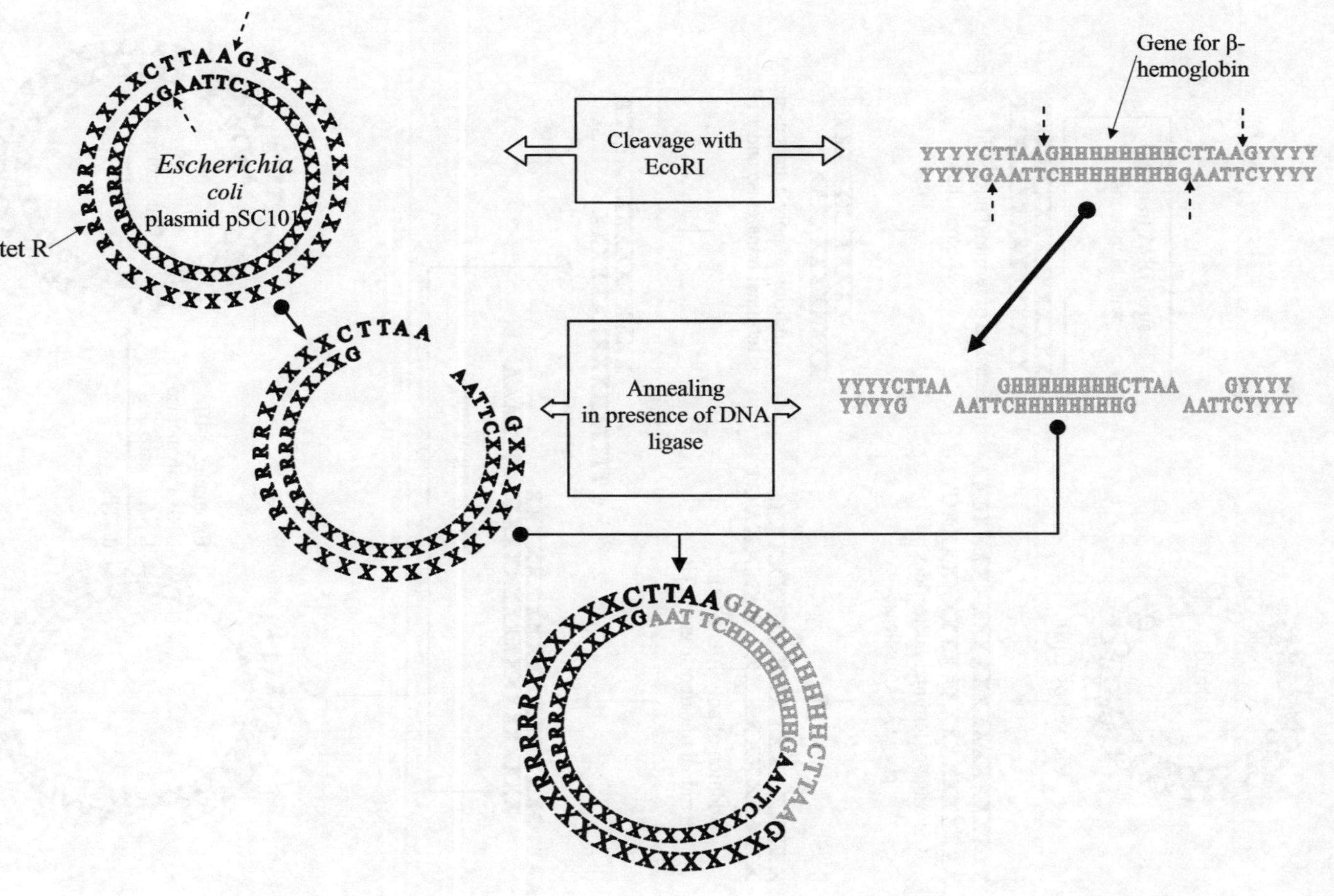

Figure. 2.18 : The construction of recombinant DNA molecule using EcoRI.

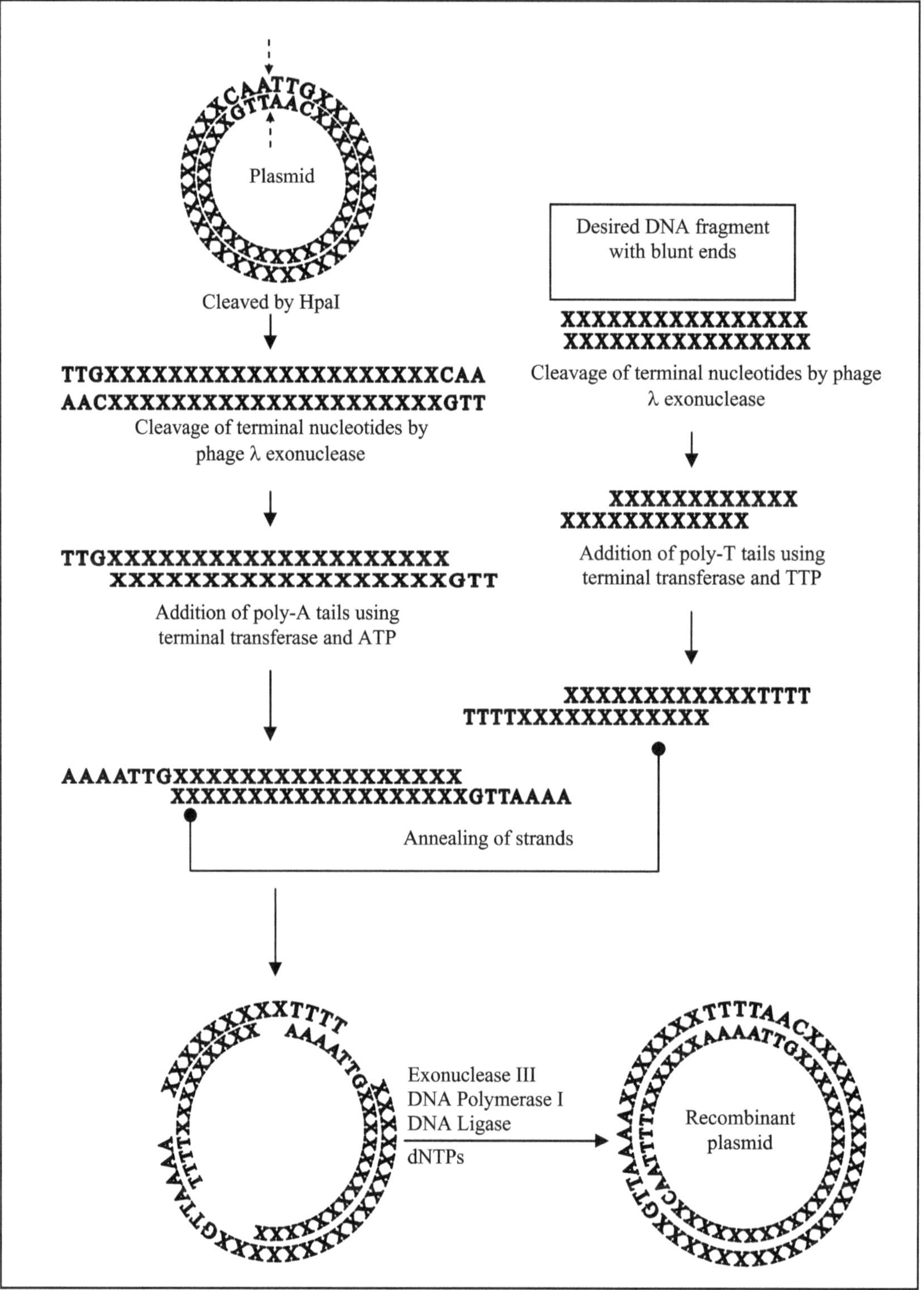

Figure. 2.19 : The construction of recombinant DNA molecule using terminal transferase procedure.

The two i.e. vector and desired gene, are treated with phage λ exonuclease to prepare substrate for terminal transferase reaction. The exonuclease removes nucleotides from the 5′ ends. The terminal transferase reaction in the presence of ATP is then used by which poly-A tails are added to 3′ OH ends of the strands of plasmid DNA. Likewise, in the presence of TTP (thymine triphosphate), poly-T tails are added to the 3′ OH ends of the strands in the foreign DNA fragments. Now the two are mixed *in vitro*. Phosphates at the end of the tail, on the 3′ ends, are removed by the action of exonuclease III. DNA polymerase I in presence of four nucleotides is used to fill the gaps. Lastly, DNA ligase is used to join the desired gene and vector.

4. Linkers and adapters

Another technique by which blunt ends can be converted to have cohesive ends involves the use of linkers and adapters.

Linkers are synthetic oligonucleotides having predetermined recognition and cleavage sites for particular RE that generates cohesive ends on cutting. The linkers are blunt ended on both the sides. First they are phosphorylated using polynucleotide kinase and then ligated to the blunt ended DNA fragment using T_4 DNA ligase. Now the desired DNA fragment is treated with the particular RE, generating the sticky cohesive ends. Now this can be ligated in the normal way with the cohesive vector. The principle is depicted schematically in Figure 2.20a and 2.20b.

5. Incompatible cohesive ends generated by use of different REs

Consider this: To obtain the desired gene, a RE is used and it generates cohesive ends. Now to insert this desired gene into a vector, suppose the vector is cleaved using a different RE (that is also not a schizomer) and that this RE also generates cohesive ends. But the result is that the generated cohesive ends both on the desired gene and the vector would be different or incompatible i.e. non-complementary. This will not allow pairing up of the bases to occur.

To overcome the incompatibility, the cohesive ends first have to be converted to blunt ends. This may be achieved by one of the two ways:

- The protruding end or the overhang may be removed using SI nuclease activity (See 2.7.1.12) or
- The single strand can be filled in complementary to the overhang, using the enzyme DNA polymerase.

Now that incompatible cohesive ends have been converted to blunt ends, the desired gene and vector can be joined using the technique of homopolymer tailing or by the use of linkers or simply by ligation using a high concentration of T_4 DNA ligase.

It is to be noted that when T_4 DNA ligase is used to join the desired gene with the vector, the ligation becomes permanent i.e. the gene cannot be recovered back ones it has performed its job (cloning or expression). But if linkers are used, then the gene can be recovered back by treating the recombinant DNA (i.e. the vector with desired gene) with the appropriate RE.

Having synthesized rDNA *in vitro*, we shall now proceed over to the next step in genetic engineering i.e. introduction of rDNA into a host cell.

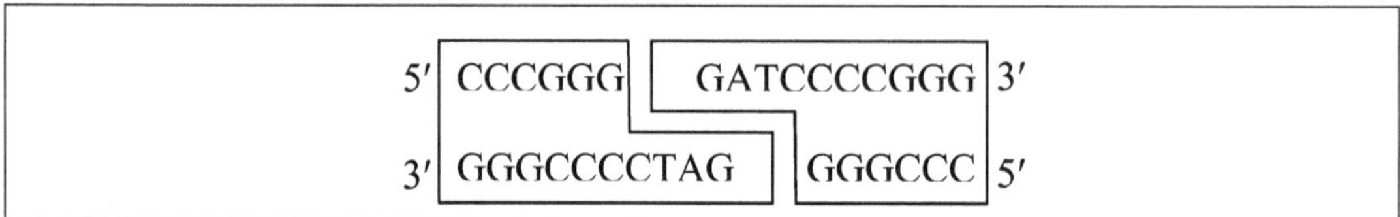

Figure. 2.20 a : A linker/adapter molecule with Bam HI sticky ends.

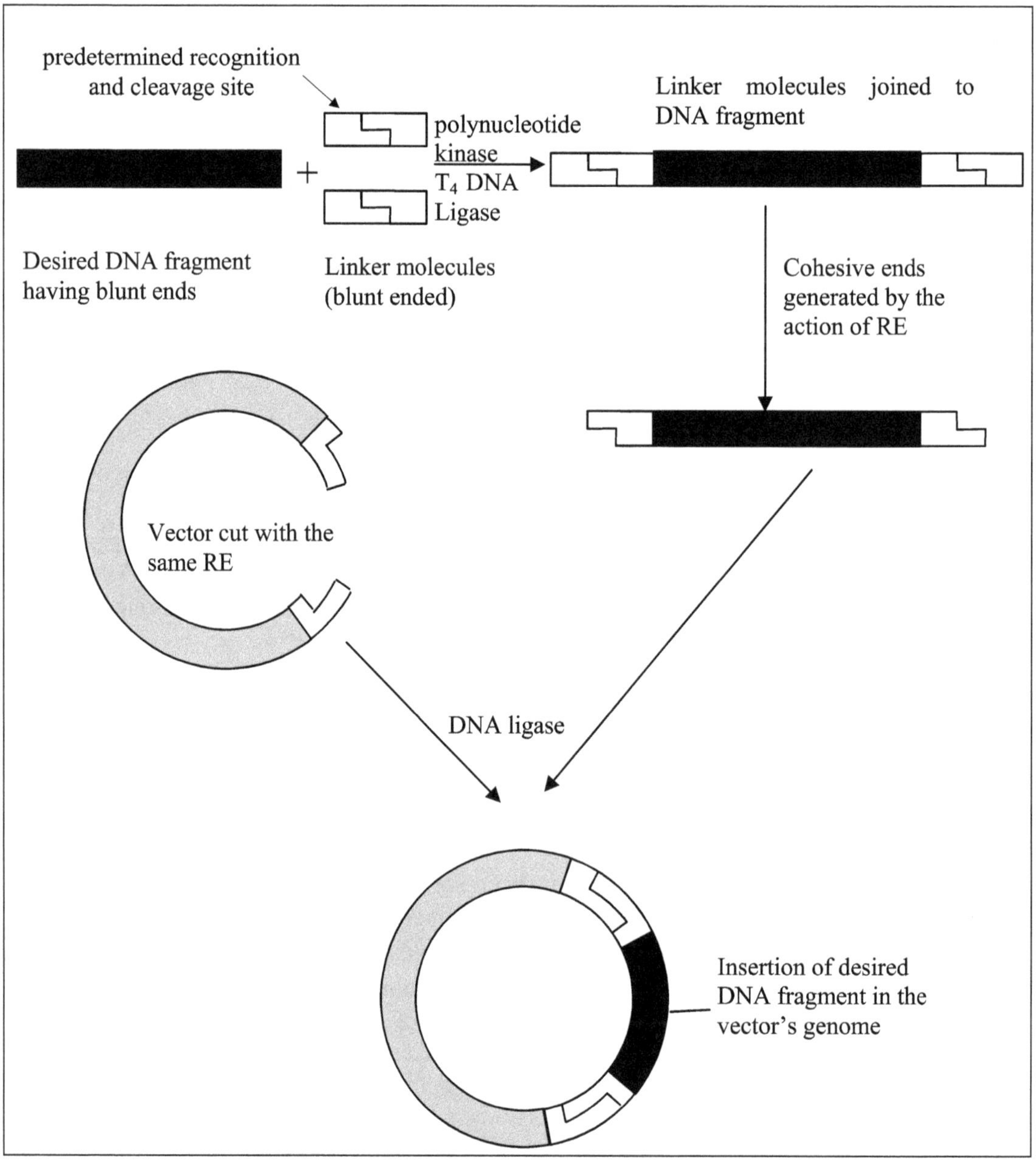

Figure. 2.20 b : Use of linkers to generate cohesive ends.

2.7.2.4 Introduction of rDNA into a host cell

If the rDNA is a plasmid then the uptake is by transformation and if the rDNA is a cosmid or a bacteriophage then introduction is achieved by transduction. Alternatively, there are other methods available, whereby the desired gene is directly introduced into the host's genome without employing the vectors.

The methods depending upon the host in which the rDNA or the gene of interest is to be introduced are considered here.

1. Bacterial cell as host

    (i) **Transformation :** Certain bacterial species like those in the genera *Streptococcus, Bacillus, Haemophilus, Neisseria* and *Rhizobium* are able to take up DNA fragments (gene of interest) spontaneously under physiological conditions. An example of one such species that is used in industrial biotechnology is *Bacilllus subtilis*. But transformation is not evident in a lot many species. *Escherichia coli*, which is most frequently employed as a host, has a very low success rate of getting transformed under normal physiological conditions. It, therefore, has become a routine step to physically and chemically treat the bacterial host so as to enhance its ability to take up foreign DNA. Such a treated bacterial cell is said to be 'competent'. Preparation of a competent bacterial cell involves treatment with 50 mM calcium chloride ice cold solution. This physico-chemical treatment is followed by a physical treatment involving a heat shock i.e. raising the temperature to 42°C to hasten up the movement of foreign DNA into the competent cell.

    (ii) **Conjugation :** This involves transfer of DNA (transposons or plasmids) from one bacterial cell to another bacterial cell via a conjugation tube. For this a special class of plasmids, called conjugative plasmids are employed to prepare the rDNA.

    (iii) **Transduction :** This involves transfer of rDNA from a virus (lambda) to a bacterial host cell. The rDNA of lambda virus can be packaged in the head and tail structure *in vitro*, requiring a number of proteins coded by lambda genome. Once the rDNA is packaged into the viral structure, it can be used in the normal way to infect bacterial host by transduction.

2. Plant cell as host

    (i) **Phytophagenia:** rDNA can be transferred to plant cells if viral vectors are used which infect plant cells by the natural infection process.

    (ii) **Ti plasmid:** *Agrobacterium tumefaciens*, is a soil bacterium that carries a conjugative plasmid called Ti plasmid. Ti stands for 'tumour inducing'. This bacterium, however, infects only dicotyledonous plants, inducing tumour growth, a disease called as crown gall. If the Ti plasmid is so manipulated such that its tumour inducing property is lost and if foreign DNA fragments are linked to it (to form rDNA), then the foreign gene can be transferred to a plant cell in the natural way. Thus, Ti plasmid is also sometimes referred to as 'Natural cosmetic engineer', since using this process, biological kingdom barriers are overcome.

(iii) **Protoplast:** In plant cells, it is the cellulosic cell wall that serves as a barrier to the entry of foreign DNA. Thus, if cell walls are removed either by enzymatic or mechanical methods, what we get is protoplast. Protoplast takes up the foreign DNA with relative ease. This phenomenon is known as transgenosis. Protoplasts take up a huge number of micro and macro molecules including chromosomes, nuclei, chloroplasts, virus and even bacteria. Once the genetic material is incorporated, cell wall synthesis starts spontaneously within a few hours.

(iv) **Micro projectile, gene gun or shot gun method**: In this technique the DNA sequence of interest is coated on 1-3 μm sized particles of tungsten or gold called as micro projectiles. The micro projectiles are loaded in the gene gun before a bullet called the macro projectile. High pressurized nitrogen drives the macro projectile along with micro projectile to hit the retaining plate which retains the macro projectile while the micro projectile under high explosive discharge and high velocity enter the plant cells.

(v) **Electroporation:** This technique involves the use of current of strength 300-800 volts/cm applied to plant protoplast, resulting in the formation of transient holes which reversibly alter the permeability of cell membrane. A recombinant plasmid along with the protoplast suspension can efficiently be taken up by the protoplast using this method.

(vi) **Ultrasonification:** In this method plant cells are placed in ultrasonic bath and ultrasonic pulses are delivered by the sonicator which generates high pressure and shock waves resulting in localized ruptures in membrane. If exogenous DNA is added to the suspension of plant cells, it can successfully be taken up by the cells.

(vii) **Liposomal delivery:** Nucleic acids encapsulated in liposomes can be transferred into target cells (plant protoplasts) by direct fusion with plasma membrane.

(viii) **Microlaser:** This is one more method as to how the cell wall can locally be ruptured and if the exogenous DNA is mixed with such cells, and then it can efficiently be internalized.

3. **Insects as host**

Insects (*Autographica californica*, commonly known as Alfalfa boor, a small moth) are eukaryotic and so possess cell machinery required for post-translational modification like glycosylation of expressed proteins. The vector that is used is Multiple Nuclear Polyhedrosis Virus, a Baculovirus, which is a class of insect virus. The desired gene is combined with the virus *in vivo* through homologous recombination, unlike *in vitro* method resulting in recombinant DNA. The homologous DNA is transferred to the insect in just the natural way as the virus infects the insect.

4. **Animal cells as host**

(i) **Viral infection :** This is self explanatory. However, it is to be noted that it is the retroviral vectors that are gaining popularity to genetically engineer all sorts of animal cells.

(ii) Microinjection : This involves injecting the exogenous DNA into the nucleus of an egg cell. This method is practically feasible with plant and animal cells since they have a larger size in comparison to bacteria. The method uses a blunt pipette onto which the egg is held by suction and at the other end a very fine needle with a small syringe filled with 50-200 gene copies, is directed to the nucleus of the fixed cell and the exogenous DNA is injected. See Figure 2.21. This figure has become a landmark in animal gene technology.

It is to be noted that the gene transfer either direct gene, rDNA, or homologous DNA is not efficiently transferred to the host cells. The success rate of the transformants is very low. The transferred foreign DNA (as it is called for bacterial hosts) or exogenous DNA (as it is called for plant and animal hosts) may be hydrolyzed by the host nucleases or may not necessarily reach and integrate with host's genome. Therefore after having introduced the gene of interest, the culture has to be screened and the successful transformants have to be identified and isolated.

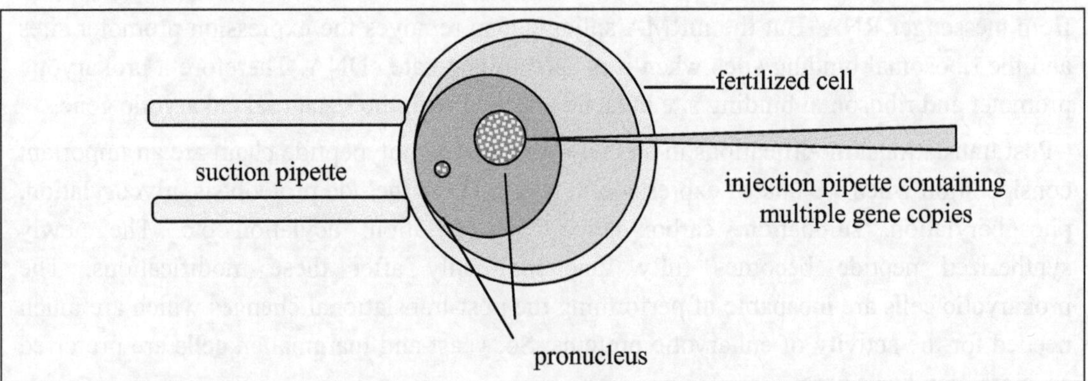

Figure. 2.21 Injecting foreign DNA into fertilized cell.

2.7.2.5 Identification and isolation of transformed cells

The transformed cells are identified on the basis of some selective property that has been acquired by the transformed cells. This selective property comes as a 'marker' along with the vector. Most frequently markers coding for specific antibiotic resistance are used. The second method that is employed uses recipient cells with specific growth deficiencies (auxotroph) and vectors carrying genes which overcome such deficiencies. Some traits exhibited by vector genes useful for identifying the transformed host cells (i.e. transformants) include:

- Resistance against antibiotics, heavy metals
- Production of antibiotics, bacteriocins, enterotoxins, H_2S
- Metabolism/ degradation of aromatic compounds, sugars, haemoglobin
- Induction of plant tumour.

2.7.2.6   Gene cloning

Having introduced the rDNA and identified the transformed host cells, the next objective is to have the gene (rDNA = gene+vector) replicated to produce many identical copies of itself or what is called as 'clone'. It is to be noted here that rDNA technology and cloning are popularly used to mean one and the same thing but in fact they are different. Construction of rDNA is a prior step, cloning is a subsequent step.

2.7.2.7   Expression

This involves transcription and translation of the information encoded by the desired gene using the host cell's machinery.

Eukaryotic genes consist of non-coding (introns) and coding (exons) elements, while the prokaryotic genes are composed entirely of exons. Therefore, if a eukaryotic gene has been introduced for expression into a prokaryotic cell, then its source should be cDNA generated from messenger RNA. But the mRMA splicing also removes the expression promoter sites and the ribosomal binding sites when it is used to generate cDNA. Therefore a prokaryotic promoter and ribosomal binding site must be attached with the structural eukaryotic gene.

Post translational modifications in a newly synthesized polypeptide chain are an important consideration when we talk of expression in hosts. These include proteolysis, glycosylation, phosphorylation, sulphation, carboxylation, hydroxylation, acylation etc. The newly synthesized peptide becomes fully functional only after these modifications. The prokaryotic cells are incapable of performing the post-translational changes which are much needed for the activity of eukaryotic proteins. So, yeast and mammalian cells are preferred as expression hosts here.

Further, the folding of proteins, to give them their tertiary and quaternary structure, is also different from the folding of proteins as it occurs in eukaryotes.

Once the gene has been expressed, its product has to be recovered from the culture. It is easier and advantageous to recover the product from periplasmic space than from the cytoplasm by lysing the cell or even from the culture broth. The movement of the product (secretion) to the periplasmic space can be achieved by engineering the gene by rDNA technique to enable its product to be secreted in the periplasmic space. This involves synthesizing and attaching a 'signal sequence' coding for a 'leader' peptide of 15-30 amino acids at the N-terminal of the gene. The leader peptide guides the movement (i.e., translocation) of the product to be secreted in the periplasmic space.

2.7.3   Typical examples of production by recombinant DNA technology

The two most commonly noted examples of the products of recombinant technology include: human insulin and the human growth hormone. A brief outline of the production of these two products is considered below.

2.7.3.1   Human insulin

The source of DNA for production of human insulin is synthetic DNA. The structural gene for human insulin consists of 1430 nucleotides. Insulin consists of two chains, chain A is made up of 21 amino acids and chain B is made up of 30 amino acids, the two chains are held together by formation of sulphide bonds between them. Note that for a total of 51 amino acids, the number of nucleotide bases required is 3×51=153 only. Using rDNA technology, the two chains are produced separately. Based on the knowledge of complete sequence of amino acids in chain A and chain B, set of appropriate oligonucleotides are linked to synthesize DNA. The synthesized DNAs for chains A and B are inserted into a vector pBR 322. The foreign DNA is inserted and ligated at the end of the gene lac Z, which is controlled by the lac promoter in the plasmid pBR 322. Also, a codon for the amino acid methionine is fused in between the gene lac Z and DNA coding for chain A. Likewise, the synthetic DNA encoding the information for chain B is inserted and ligated in the vector pBR 322 at the end of gene lac Z. Also a code for methionine is built in at the fusion point of the two. The rDNA is then introduced into the host i.e. *Escherichia coli* (see Figure 2.22).

The lac promoter controls the expression of lac Z. The product of lac Z gene is ß-galactosidase. Thus, if the code for methionine and DNA coding for chain A or B is fused to gene lac Z, what we get is a fused product i.e. ß-galactosidase-methionine-insulin chain A or B. The reason why such a fusion product needs to be expressed is that the fusion product is prevented from being proteolyzed else the chains expressed above undergo proteolysis.

Having expressed the fusion product, the next step is treatment with cyanogen bromide. This agent has the characteristic property of cleaving after a methionine bond. The chain A or B do not have methionine in their structure. Thus by treating with cyanogen bromide, the chains are cleaved off from ß-galactosidase-methionine. The two chains are then isolated and mixed together to allow the sulphide bond formation which is spontaneous.

2.7.3.2   Human growth hormone [HGH]

The source of DNA used is cDNA, reverse transcribed from messenger RNA derived from the human pituitary. The cDNA is further processed with RE to remove the information required for its transport in the human cell. But this step causes the loss of some of the amino-acids essential for the activity of the hormone. The information thus lost is chemically synthesized and fused to the cDNA. Further, a nucleotide sequence coding for a leader peptide is fused to the cDNA. This construct is then inserted at the end of a strong promoter, into a bacterial vector. The host employed for expression is *Escherichia coli*. The hGH gene is expressed producing the hormone attached with the leader peptide which helps in translocating the hormone to the periplasmic space. Simultaneously, the leader peptide is cleaved off and the hormone is set free. The translocation of hormone to the periplasmic space makes purification of hGH easier and cheaper.

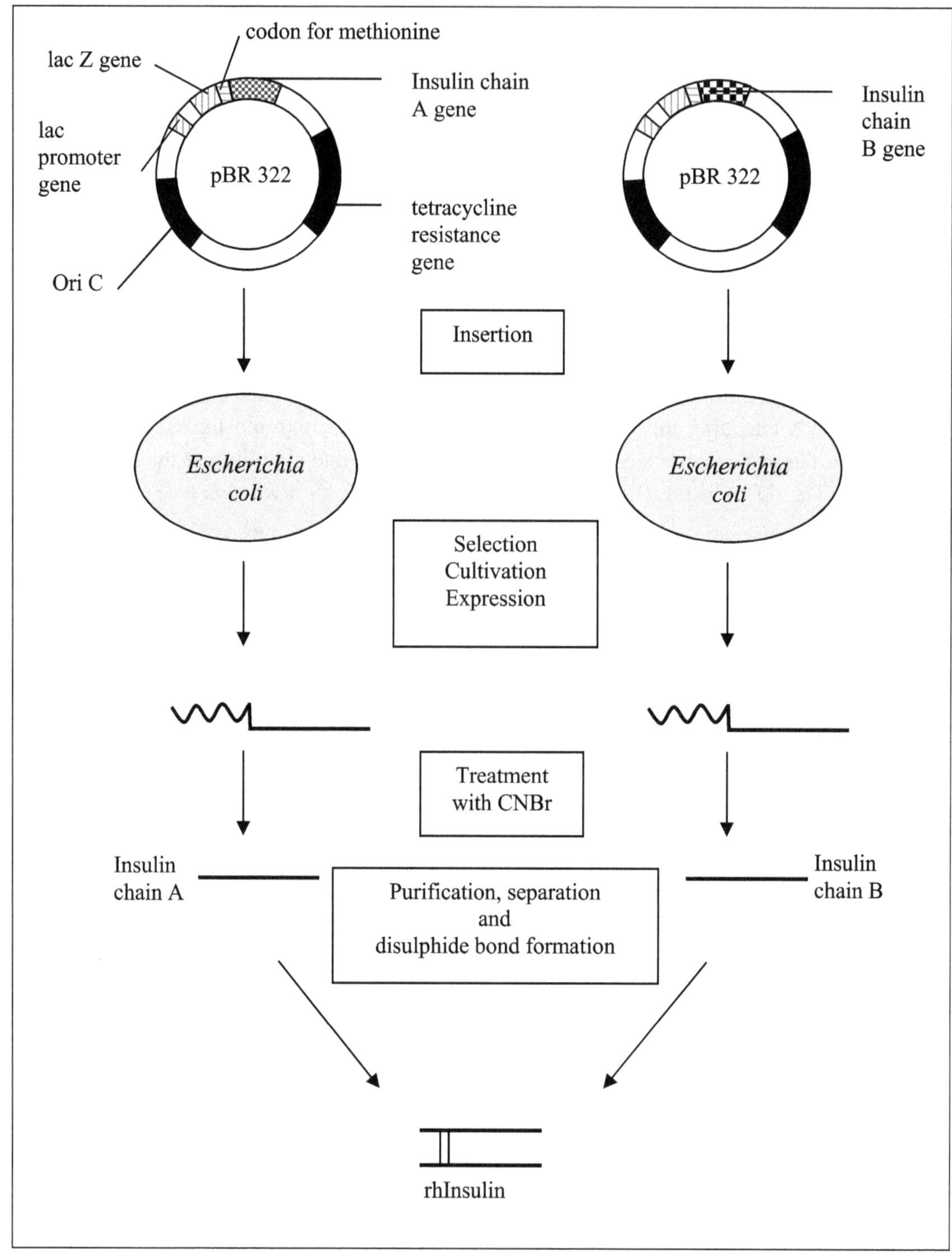

Figure. 2.22 : Outline of the production of recombinant human insulin.

2.7.4 Commercial products

Organic chemistry and medicinal chemistry are already the established disciplines in the field of drug discovery and development. Since the last three decades, through the efforts made in modern biotechnology i.e. genetic engineering, over 75 products have been developed and are marketed for therapeutic use. The first biotech product based on recombinant DNA technology was human insulin which was marketed in the year 1982 by the company Elli Lilly by the name Humulin. Most of the genetically engineered molecules are glycosylated proteins and include biological response modifiers (BRM), haemopoietic growth factors (HGF), enzymes, hormones, monoclonal antibodies (MAb) and vaccines. A list of such products is presented in the Table 2.5. Several hundred biotech based products are in the pipeline in the developmental/ approval stage. Some commercial products are discussed below.

2.7.4.1 Alteplase

This is recombinant tissue-type plasminogen activator. A thrombus is deposition of fibrin and platelets resulting in clotting of blood in vasculature. This leads to thromboembolic diseases that are responsible for number of deaths. The condition can be managed by having the fibrin digested. Physiologically, the protective, fibrin-digestive action begins with plasminogen, which is an inactive zymogen or proenzyme that is activated to plasmin. This active moiety, plasmin, acts upon the insoluble fibrin matrix of a thrombus and digests it to yield soluble fibrin degradation products. The very first step i.e. activation of plasminogen to plasmin (see Figure 2.23) is physiologically brought about by plasminogen activators like tissue-type plasminogen activator (t-PA) or urokinase-type plasminogen activator (u-PA).

The tissue-type plasminogen activator (t-PA) has a fibrin-specific plasminogen activation which results in fewer systemic side effects and also in comparison to other non-endogenous plasminogen activators like streptokinase; t-PA does not have allergic or hypotensive effects.

(a) *Structure :* t-PA consists of a single chain made up of 527 amino acids and has a molecular weight of 64 kDa, (6%-8%) of which is contributed by attached carbohydrates.

    In the structure shown in Figure 2.24, seventeen disulphide bridges could be seen between the sulphur containing amino acid cysteine. Further at amino acid numbers 117, 184 and 448 (asparagine, represented by N) there is presence of a carbohydrate moiety i.e. it is glycosylated by N-linkage. Two forms of t-PA exist, type I and type II. Type I is glycosylated at 117,184 and 448. While type II is glycosylated at 117 and 448 only. Glycosylation is a post-translation modification of the protein which influences the activity and pharmacokinetics of t-PA.

The recombinant t-PA or rt-PA or commonly known as alteplase is identical to the endogenously produced human t-PA. The recombinant enzyme was first produced using *Escherichia coli* as the host. But this host lacked the ability to glycosylate and also lacked the ability to form the correct 3-dimensional t-PA structure. So, a mammalian host was selected, the choice being Chinese hamster ovary (CHO) cell line, which is the most commonly employed mammalian host cell line. In CHO cell line, glycosylation, disulphide bond formation and proper folding as desired is achieved. Initially roller bottle process was used for producing rt-PA, however subsequently rt-PA is being produced by a large scale suspension culture process.

Table 2.5 : Some marketed products produced using recombinant technique.

PRODUCT[b]	CLASS [a]	PRODUCT[b]	CLASS [a]
Abciximab	MAb	Insulin (3)	Hormone
Acelluvax Vaccine	Interferon alpha (5)	BRM	
Aldesleukin	BRM	Interferon alpha con	BRM
Alteplase Enzyme	Interferon beta (2)	BRM	
Antihemophilic Factor VIII (2)	Enzyme	Interferon gamma	BRM
Basiliximab	MAb	Lenograstim	HGF
Capromab penetrate	MAb/dx	Liposomal agents (5)	Liposome
Daclizumab	MAb	Lyme disease	Vaccine
Denileukin diftitox	BRM	Molgramostism	HGF
Dnase Enzyme	Muromonab-CD3	MAb	
Entanercept	BRM	Nartograstism	HGF
Epoetin alfa (6)	HGF	Nefotumomab	MAb/dx
Eptifibatide	Enzyme	Oprelvekin	BRM
Filgrastim HGF	Panorex	MAb	
Factor VII Enzyme	Palivizumab	MAb	
Factor IX Enzyme	Peg-interferon (2)	BRM	
Follitropin (2) delivery	Hormone	Polymer- BCNU	Drug
Ganirelix Peptide	Rituximab	MAb	
Gemtuzumab	MAb	Retiplase	Enzyme
Glatiramer	BRM	Sargramostim	HGF
Glucagon Hormone	Somatropin (6)	Hormone	
Growth Hormone Releasing Hormone	Hormone Pendetide	Satumomab	MAb/dx
Hepatitis B Vaccine (2)	Vaccine	Tenecteplase	Enzyme
Hyaluronate Acid Materials	Surgical adjuncts	Tirobifan	Enzyme
Imiglucerase	Enzyme	Trastuzumab	MAb
Imiciromab Penetrate	Mab/dx	Vitravine	NA
Infliximab MAb			

a.   Abbreviations include: MAb= Monoclonal Antibody; MAb/dx=MAb for diagnosis, HGF=Hematopoietic Growth Factor; BRM=Biological Response Modifier, NA=Nucleic Acid

b. The number in parentheses indicates the approximate number of products of the same molecule manufactured by different companies.

2.   Product information

- rt-PA i.e. Alteplase is available by the brand names:
  Activase from Genentech INC and
  Actilyse from Boehring
- Supplied as sterile, white to off-white lyophilized powder for reconstitution.
- Insoluble in water, so arginine is added in the formulation to enhance its solubility by complexation.
- Storage: $2^{\circ}C$-$8^{\circ}C$ never to exceed $30^{\circ}C$ and protected from light.

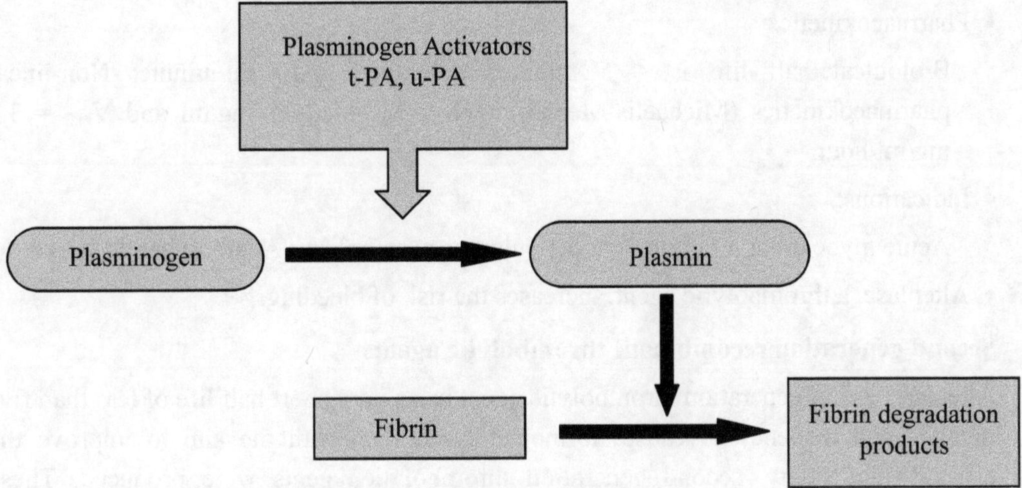

Figure. 2.23 : The fibrinolytic pathway.

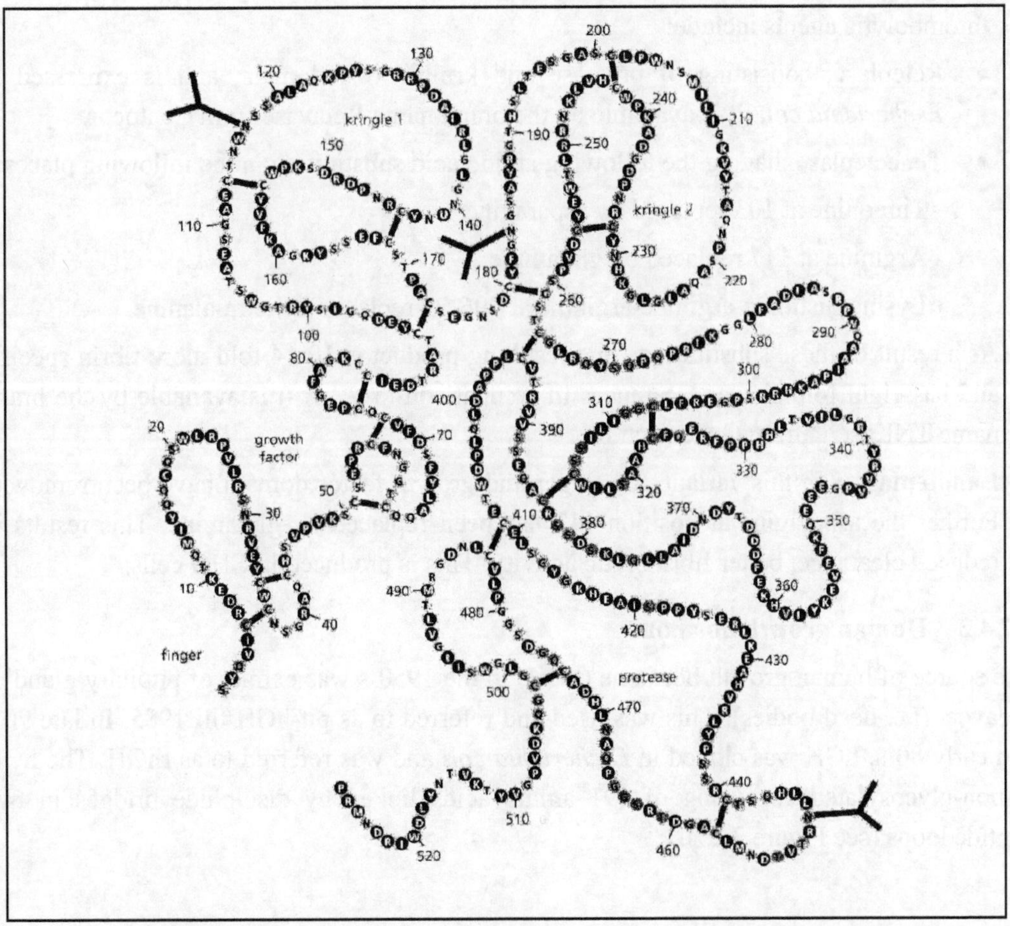

Figure. 2.24 : Primary structure of tissue – plasminogen activator.

- Pharmacokinetics:

  Biological half life is < 5 minute; Clearance: 6-20 ml/minute; Non-linear pharmacokinetics (Michaelis-Menten) with a K_m = 12-15 mg/ml and V_{max} = 3.7 mg/ml/hour.

- Indications:

  Acute myocardial infarction; Acute pulmonary embolism; Acute ischemic stroke.

- Alteplase, a thrombolytic agent, increases the risk of bleeding.

3. Second generation recombinant thrombolytic agents

Alteplase, a first generation thrombolytic agent has a very short half life of less than five minutes and tendency to cause fibrinogenolysis, thus with the aim to improve the efficacy and safety, second generation thrombolytic agents were produced. These variants of t-PA were genetically engineered to have domain deletions, glycosylation changes and site-directed amino acid substitutions. The second generation recombinant thrombolytic agents include:

- Reteplase, consisting of protease and kringle two domain, and is expressed in *Escherichia coli*. It is available by the brand name Retavase from Centocor.

- Tenecteplase, having the following amino acid substitution in the following places:

  - Threonine at 103 replaced by asparagine.

  - Arginine at 117 replaced by glutamine.

  - Lysine-histidine-arginine-arginine at 496-99 replaced by tetra-alanine.

As a result of these substitutions, the resulting product is 10-14 fold more fibrin specific and has eight fold slower clearances in comparison to t-PA. It is available by the brand name TNKase™ from Genentech Inc.

4. Lanoteplase : In this variant, the finger and growth factor domain have been removed. Further the asparagine at position 117 has been replaced by glutamine. This results in reduced clearance, better fibrinolytic activity. This is produced in CHO cells.

2.7.4.2 Human growth hormone

The source of human growth hormone (hGH) in the 1950's was extract of pituitary gland of cadavers (i.e. dead bodies). This was used and referred to as pit-hGH till 1985. In late 70's and early 80's, hGH was cloned in *Escherichia coli* and was referred to as rhGH. The hGH is non-glycosylated consisting of 191 amino acids linked by disulphide bridges in two peptide loops (see Figure 2.25).

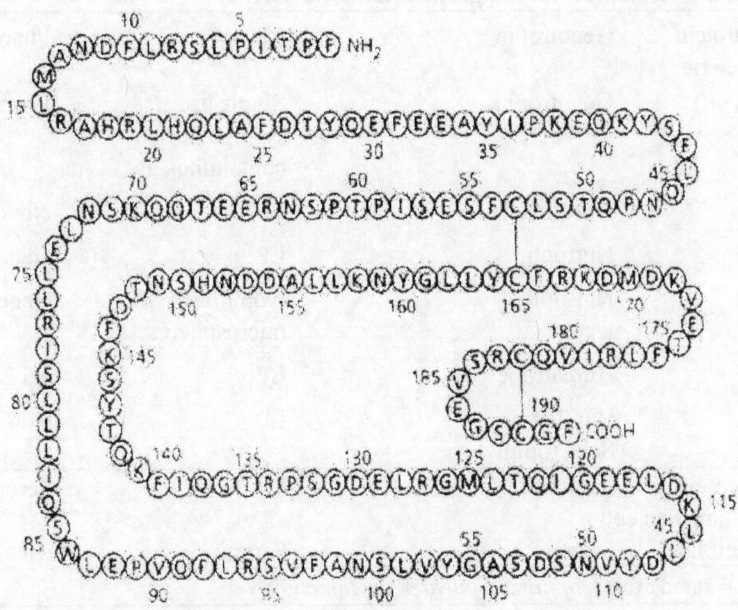

Figure. 2.25 : Human growth hormone: Primary structure.

Nowadays rhGH is produced both in *Escherichia coli* and in mammalian cell lines. If *Escherichia coli* is used as the host, rhGH has got to be released from the periplasmic space by osmotic shock and then recovered. If mammalian cell line is used, then it is secreted directly out side the cell and has to be recovered from the culture medium.

A compilation of the hGH preparations is presented in Table 2.6

Table 2.6 : Marketed preparations of different types of human growth hormone

Source Brand	Product	Supplier	Name Form
Pituitary-derived human growth hormone (pit-hGH)			
Cadaver pituitaries	Crescormon	Discontinued from market	Pharmacia (formerly KabiVitrum AB)
	Nanormon	"	Novo Nordisk
Somatrem, methionyl human growth hormone (met-rhGH)			
Recombinant protein produced in bacteria (*Escherichia coli*)	Somatonorm	Discontinued from market	Pharmacia (formerly KabiVitrum AB)
	Protopin	LP	Genetech, Inc.

Table 2.6 Contd...

Somatropin, natural sequence human growth hormone (rhGH)

Recombinant protein produced in bacteria	Genotropin	LP	Pharmacia Corp
(*Escherichia coli*)	Genotropin Miniquick	Single use syringe devise containing LP	Pharmacia Corp
	Norditropin	LP	Novo Nordisk
	Nutropin	LP	Genetech Inc.
	Nutropin Depot	lyophilized SR microspheres	Genetech Inc.
	Humatrope	LP	Eli Lilly & Co.
	Bio-Tropin Sci Tropin	LP	Bio-Technology General Corp.
Recombinant protein produced in mammalian cell	Serostim	LP	Serono
(C127 mouse cell line)	Saizen -		Serono

Abbreviation LP stands for: *lyophilized powder for injection*

2.7.4.3   Interferon

Interferon is a substance that interferes with viron replication. Currently the interferons are classified on the basis of physio chemical and antigenic properties into three classes: interferon alpha (IFNα) primarily produced by leukocytes; interferon beta (IFNβ) primarily produced by fibroblasts and interferon gamma (IFNγ) produced by lymphocytes.

At least Fourteen different sub-types of natural IFNα are known. Prior to the advent of rDNA technology, until 1980, IFNα was obtained from the pooled units of human leukocytes that had been induced to release interferon by incomplete injection with avian Sendai virus. The mixture consisting of sub types of IFNα was purified and was designated as IFNα, using immunoaffinity chromatography with monoclonal antibodies.

1. **Interferon α (IFNα) :** Using rDNA technology, a single IFNα sub type can be produced. Three recombinant versions of IFNα are presently marketed which include the following:

    (i) **Roferon-A®** : This is recombinant IFNα-2a which is commercially available from L'Hoffman LaRoche. It consists of 165 amino acids having lysine at position 23. The source of the DNA sequence encoding for the protein, is cDNA obtained from the mRNA from IFN producing human leukocytes. It is inserted into the vector pBR322 and introduced into the genetically engineered *Escherichia coli*, most commonly in the β-lactamase gene at Pst site. Since it is expressed in *Escherichia coli*, it is a non-glycosylated protein. Purification is done by affinity chromatography with the use of murine monoclonal antibodies.

    - IFNα-2a is also commercially available as Wellferone from Glaxo-Wellcome group and as Shanferone from Shantha Biotech Ltd.

(ii) Intron-A® : This is recombinant IFN-α2b which is commercially available from Schering-Plough. It consists of 165 amino acids with arginine at position 23. The source of its DNA sequence is cDNA obtained from mRNA by the action of reverse transcriptase. This is also recombined with a vector like pBR322 and introduced for expression into genetically engineered *Escherichia coli*. The purification is done by proprietary methods.

(iii) Infergen® : This is recombinant interferon alfacon-1. The source of its genetic sequence is synthetic DNA which is based upon the comparative amino acid sequences of the several natural interferons and sub types or i.e. to say IFN- α consensus. It is a construct of 166 amino acids having the most commonly occurring amino acids in the appropriate position as they occur in sub types. Its amino acid sequence is 88% homologous with IFN- α2 and 30% homologous with IFN- β. The synthetic DNA is likewise introduced into genetically engineered *Escherichia coli* for expression.

The principle use of interferon α is immunostimulation and different forms of IFN- α (as mentioned above) are indicated to fight out a variety of cancers (such as hairy cell leukemia, genital warts, AIDS-associated Kaposi's sarcoma, bladder carcinomamalignant melanoma etc.) and viral infections (such as laryngeal papilloma, mycosis, fungoides etc.).

2. Interferon β (IFN-β): IFN- β has both antiviral and immuno-regulatory effects. Further, the activities of IFN- β are species specific. Two forms of recombinant IFN-β are commercially available.

(i) Betaserone: This is rIFN- β-1b. The source of its gene is the human fibroblasts. But the obtained gene is altered so as to substitute a serine codon for the cysteine codon found at position 17. The protein is non-glycosylated and so is introduced into genetically engineered *Escherichia coli* for its expression. It is indicated foe patients with relapsing-remitting multiple sclerosis (MS)

(ii) Aronex: This is rIFN- β-1a. It is approved for treating relapsing forms of MS.

3. Interferon γ (IFN- γ): IFN- γ is produced by antigen activated T-cells. After its release it binds to surface receptors of resting macrophages and monocytes to induce them, resulting in enhanced phagocytic activity of these cells. The immuno-stimulation of phagocytic cells brought about by the IFN- γ is much more pronounced than that of IFN- α or β however the antiviral activity of IFN- γ is less than that exerted by IFN- α or β.

The human IFN- γ is a protein consisting of 143 amino acids, having a little homology in sequence compared to either IFN- α or β.

- **Actimmune:** This is rIFN- γ-1b consisting of 140 amino acids produced in *Escherichia coli* and purified by column chromatography. It is indicated for treating infections associated with chronic granulomatous disease and for treatment of osteoporosis.

2.7.4.4 Vaccines

Conventionally vaccines are defined as immunological preparations which are:

- suspensions of micro organisms either whole bacteria or virus. The micro organisms may be dead (killed vaccine) or live but weakened (attenuated vaccines); or

- toxoids i.e. a toxin that has been physically (heat) or chemically (formaldehyde) treated in such a way so as to remove the toxic properties while keeping the immunogenic property intact.

With the advent of genetic engineering it is now becoming possible to produce modern vaccines. The various biotechnology based modern vaccines include:

1. **Genetically attenuated live vaccines** : This involves deletion of the pathogenesis-causing genes from the pathogen to make it nonvirulent, example: genetically attenuated virulent strain CVD 103.

2. **Gene cloning or live vectors:** - This involves inserting of genes encoding for surface antigens or the particular part of an antigen (i.e. epitope) from a pathogenic microorganism into a harmless, non-pathogenic or attenuated or live microorganism. The principle is illustrated in Figure 2.26.

   The two most commonly used harmless microorganisms for expression of the foreign antigen include vaccinia virus and *Salmonella* species. Examples of this strategy include: the expression of antigen of HIV, rabies, influenza etc. on Vaccinia; and expression of antigen of *Bordetella pertusis*, Hepatitis Bordetella virus, *Plasmodium* species, *Shigella* etc. using *Salmonella* species.

3. **Genetic detoxification by site directed mutagenesis:** - Conventionally toxoids are produced by subjecting the native toxin to physical or chemical treatment so as to nullify its toxic properties while keeping the antigenic (immunogenic) properties intact. This harsh treatment generally reduces the immunogenecity, as well, of the toxoid to some extent and there always exists a possibility of reversal of the toxoid to a biologically active toxin. Therefore, using the genetic detoxification approach, the gene encoding for the toxic part is specifically mutated so as to produce a mutant which expresses a non-toxic protein and at the same time it is highly immunogenic.

   Applicable example: *Bordetella pertusis* containing altered gene, abolishing the toxicity of the protein (toxin P). Other candidates include toxin producing bacteria like diphtheria, tetanus, cholera.

4. **Expression of antigens in host cells:** - This involves formation of rDNA i.e. a plasmid with an antigen encoding gene incorporated in its DNA. This is then used to transform a host cell like baker's yeast (*Saccharomyces cerevisiae*) or mammalian cell (CHO cell) wherein the antigen (protein) shall be expressed.

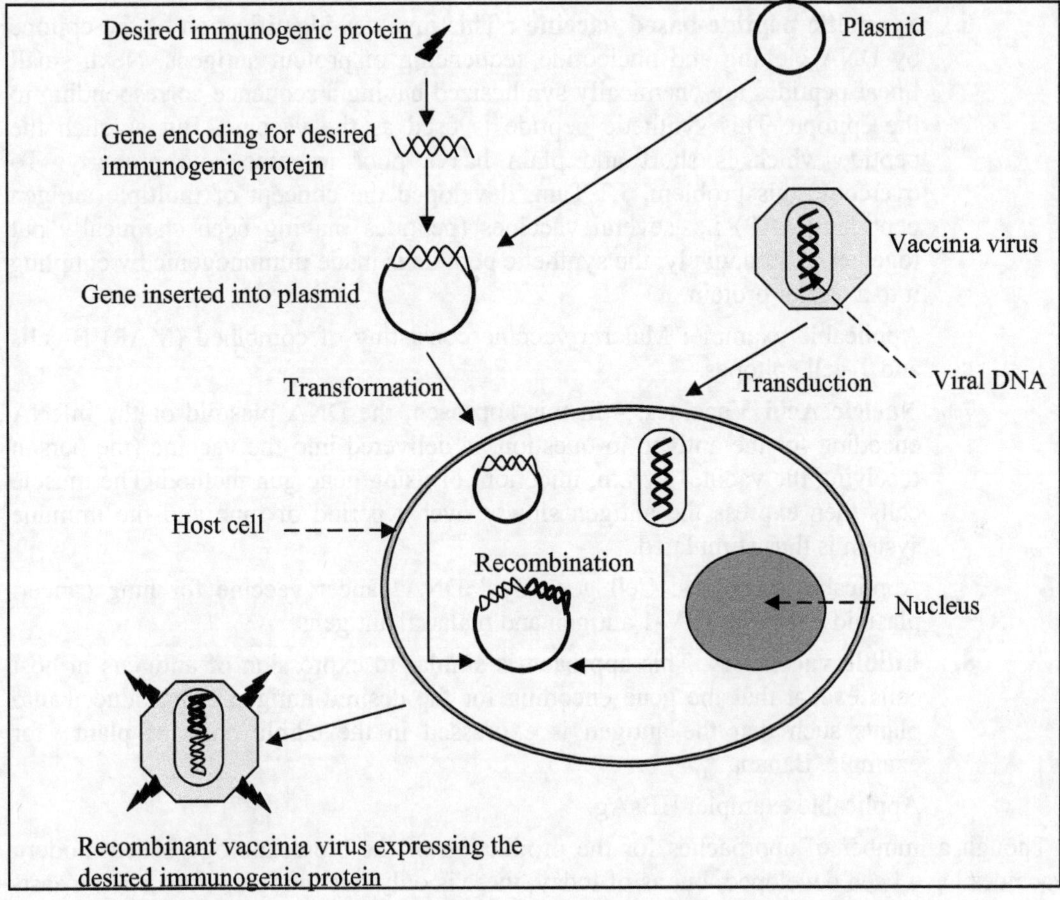

Figure. 2.26 : Vaccine production: Strategy involving live vectors.

This method is used to produce hepatitis B surface antigen (HBsAg). In both the expression systems, HBsAg is found as particulate aggregates of 22nm size. These aggregates are identical to the aggregates secreted by the HBV in the plasma. The yeast derived vaccine is commercially available world wide by the brand name Recombivax HB from Merck and Shanvac from Shantha Biotech, Hyderabad, India.

5. **Anti-idiotype antibody vaccine**: - Antibodies are highly specific and recognize an antigen or its critical part i.e. epitope of an immunogen. The monoclonal antibody generated against a particular epitope, is referred to as an idiotype antibody. The approach here is to generate monoclonal antibody against the idiotype antibody i.e. anti-idiotype antibody. This anti-idiotype antibody, thus, has a 3-D structure similar to the structure of the epitope of the immunogen. Thus anti-idiotype antibody can be used as a vaccine.

Applicable example: Vaccine consisting of anti-idiotype antibody resembling lipopolysaccharide from *Pseudomonas Aeruginosa* and vaccine against *Strepocuccus pneumoniae* based on phosphorylcholine-mimicking antibodies.

6. **Synthetic peptide-based vaccine** : This involves identification of the epitope by DNA cloning and nucleotide sequencing of protein antigens. Next, small linear peptides are chemically synthesized having a sequence corresponding to the epitope. This synthetic peptide is used as the vaccine. But as such the peptide which is short and plain has a poor immunogenic property. To overcome this problem, J.T.Tam, developed the concept of multiple antigen peptides (MAP) i.e. several vaccines (peptides) having been chemically put together. Alternatively, the synthetic peptide is made immunogenic by coupling it to a carrier protein.

    Applicable example: Malaria vaccine consisting of combined (MAP) B-cells and T-cell epitopes.

7. **Nucleic Acid Vaccines**: - In this approach, the DNA plasmid or the mRNA encoding for the antigen in question, is delivered into the vaccine (the person receiving the vacine) by i.m. injection; or using gene gun method. The muscle cells then express the antigen slowly over a period of time and the immune system is thus stimulated.

    Applicable examples: Cell gene Sys' DNA cancer vaccine for lung cancer; plasmid encoding HIV-1 antigen and malarial antigens.

8. **Edible vaccines:** - This approach is similar to expression of antigens in host cells except that the gene encoding for the desired antigen is introduced into plants such that the antigen is expressed in the edible parts of plant for example. Banana.

    Applicable example: HBsAg.

Though a number of approaches for the production of the biotechnology based modern vaccines have been developed, but as of today, there is only one modern vaccine i.e. Yeast-derived HBsAg vaccine, which is commercially available. The second modern vaccine likely to be commercialized is the improved acellular pertusis vaccine (genetically detoxified). So there is reluctance to give up the conventional vaccines and at the same time there is concern over the safety of the modern biotech based vaccines.

2.7.4.5   Haematopoietic Growth Factors

Haematopoietic refers to blood. Blood cells carry out vital physiological functions like transportation of oxygen, development of immunity and clotting of blood. Haematopoiesis is the process of production and maturation of blood cells. The immature precursor cells i.e. haematopoietic stem cells found in the bone marrow undergo maturation i.e. proliferation and differentiation in the presence of a number of agents including most importantly the haematopoietic growth factors, eventually being converted into differentiated functional hematopoietic cells, such as erythrocytes, platelets, macrophages, granulocytes etc.

The general outline for the production of rhHGF includes: Identification and isolation of gene in question, inserting it into a plasmid and introduction into a host for expression.

Commercially available Haematopoietic Growth Factors are presented in Table 2.7.

Table 2.7 : The commercially available Haematopoietic Growth Factors.

Generic name	Brand name	Company	Source of gene	Expression system	Applications
			Granulocyte-Colony stimulating Factor (rhG-CSF)		
Filgrastim	Neupogen Granulokine Gran	Amgen Dompe Kirin	bladder carcinoma cell line	*E. coli*	neutropenia, aplastic anaemia, hematopoesis after bone marrow transplantation, prevention of infections in AIDS
Lenograstim	Granulocyte Euprotin Neutrogin	Rhone-Poulenc Amrad Chugai	squamous carcinoma cell line	Chinese hamster ovary cell line	"
			Granulocyte Macrophage-Colony Stimulating Factor (rhGM-CSF)		
Molgramostim	Leucomax Mielogen	Schering Schering-Plough	human monocyte cell line	*E. coli*	hematopoiesis after chemotherapy & bone marrow transplantation, mobilization of cells for transplantation stem
Sargramostim	Leukine	Immunex Wyeth-Ayerst	mouse T lymphoma cell line	*S. cervisiae*	"
			Erythropoietin (rhEPO)		
Epoetin alfa	Epogen Procrit Eprex Espo	Amgen Ortho " Kirin, Japan	-	Chinese hamster ovary cell line	treatment of anemia associated with chronic renal failure, in predialysis patients, treatment of zidovudine (antiHIV) and chemotherapy induced anemia, for increasing the yield of autologous blood prior to donation
Epoetin beta	Epogin NeoRecormon	Chugai, Japan Boehringer Mannheim	-	"	"
			Stem Cell Factor (rhSCF)		
Ancestim	Stemgen	Amgen	-	*E. coli*	to increase mobilization of stem cells for autologous transplantation
			Interleukine-11 (rhIL-11)		
Oprelvekin	Neumega	Genetics Institute	-	-	prevention of thrombocytopenia

2.7.4.6 Coagulation factors

Normally it is observed that whenever there is bleeding due to tissue injury, it is arrested within a few minutes and that the blood sets into a jelly like mass. This phenomenon is termed as coagulation or clotting. Blood coagulation may be initiated by one of the two pathways i) The exposure of blood to extrinsic factors i.e. tissue factors derived from outside the plasma; ii) By the activation of intrinsic factors i.e. factors present in the plasma itself. Both the pathways converge at the point where active factor X (or Stuart factor) is formed; thereafter the pathway is common leading to the formation of fibrin. The mechanism of blood clotting is depicted in the Figure 2.27.

A total of thirteen factors are involved in the mechanism of blood clotting. A deficiency or lack of any of the factors results in the abnormality in blood coagulation. One such disorder resulting due to decreased production or absence of factor VIII is known as Haemophilia A. Patient lacking factor IX are said to have Haemophilia B. For Haemophiliacs blood transfusion temporarily provides factor VIII and stops bleeding. Now the recombinant coagulation factors are available for haemophilic patients, providing improved safety and reducing the risk of infection transmitted by transfusion.

Commercially available coagulation factors are presented in Table 2.8.

2.7.4.7 Interleukins

Cells produce molecules that are soluble mediators or glycoproteins that aid in the communication between cells. These mediator molecules are called cytokines of which 60 different types are known. Interleukins which bring about communication between WBC, especially T-cells are one such class of glycoproteins of which eighteen types are known. However, only two interleukins are produced using rDNA technology and these are commercially available. They are IL-2 and IL-18.

1. **Interleukine-2**: rhIL-2, generic name aldesleukin, is marketed by the name proleukin. This is not glycosylated hence produced in *Escherichia coli*. In contrast to the native forms, the recombinant form has a serine substituted for cystein at position 125. It binds with IL-2 surface receptors of inactivated T-cells and brings about T-cell proliferation and differentiation. It induces activation of natural killer (NK) cells and lymphokine activated killer (LAK) cells. It also brings about the release of tumor necrosis factor, IL-1, γ-interferon and GM-CSF from the cells with which it binds.

    It is used in the treatment of a variety of neoplastic diseases.

2. **Interleukin-11**: rhL-11, generic name oprelvekin, is available as Neumega. This is also non-glycosylated hence produced in *Escherichia coli*. It consists of 177 amino-acids and in contrast to the native IL-11, consisting of 178 amino-acids, rhIL-11 lacks the terminal proline at 178. It is the thrombopoietic growth factor. It stimulates proliferation of haemopoeitic stem cells, induces megakaryote maturation, resulting in increased platelet production.

    It is indicated for the prevention of severe thrombocytopenia.

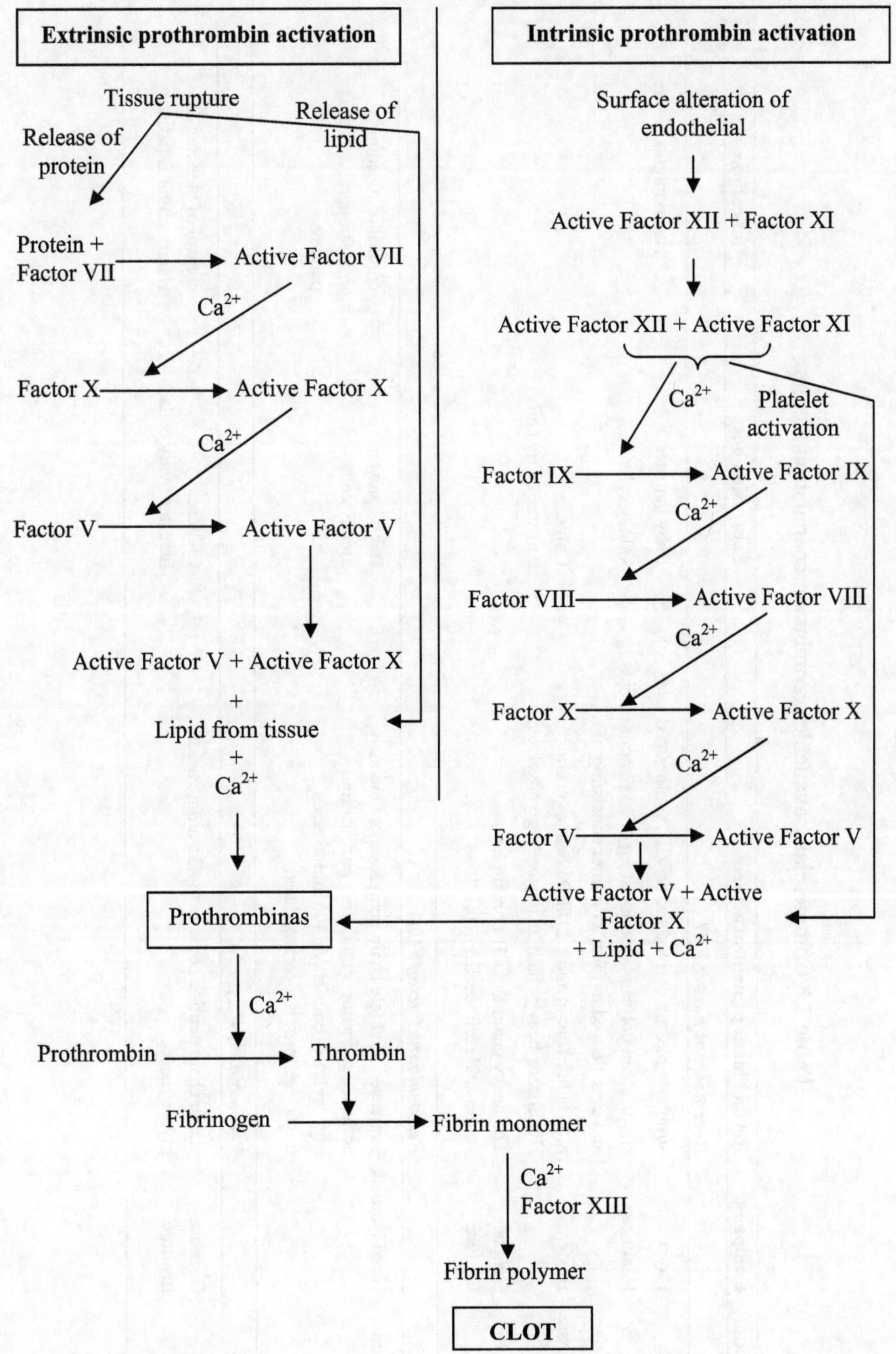

Figure. 2.27 : Process of coagulation of blood: the involvement of blood factors.

Table 2.8 Commercially available recombinant coagulation factors.

Brand name	Company	Recombinant protein structure	Expression host	Indications
		Recombinant Factor VIII		
Kogenate	Bayer Healthcare	similar to plasma derived factor VIII, synthesized as a single chain polypeptide consisting of 2332 amino acids and gets cleaved to form a dimmer consisting of	Baby hamster kidney cells	Haemophilia A
Recombinate	Baxter Healthcare	80kDa light chain and a 210kDa heterogeneous heavy chain, extensively glycosylated.	Chinese hamster ovary cells	”
Refacto	Genetics Institute	The heavy chain lacks B-domain i.e. amino acid residues from 741-1648	”	”
		Recombinant Factor VIIa		
Novoseven	Novo Nordisk	expressed as single chain peptide, spontaneously activated to Factor VIIa during purification, glycosylated, consists of 406 amino acids, the active form is γ carboxylated	Baby hamster kidney cells	control of bleeding in Haemophilia A or B patients
		Recombinant Factor IX		
BeneFix	Genetics Institute	single chain peptide consists of 415 amino acids, glycosylated.	Chinese hamster ovary cells	control of bleeding in Haemophilia B patients

2.7.5 Gene Therapy

In the previous section, biotechnology (gene) based pharmaceuticals were considered. In this section, introduction of gene(s) into cells of patients suffering from genetic disorders so as to improve the prognosis or to cure the disease i.e. gene therapy shall be considered.

In gene therapy, one of the two approaches can be adopted.

One : introduction of genes in germ cells i.e. sperm or ovum or into the omnipotent embryonic cells i.e. at 4-8 cellular stage. As a result of this, all the descendents of the person receiving the genes will have an altered genome. Also, all the cells of the individual, developing from the embryo, will carry the therapeutic gene and will pass down over to the next generations. This approach is unacceptable ethically as of today.

Two: introduction of genes in somatic cells, i.e. any of the cells of body except germ and embryonic cells. This will, therefore, affect only the engineered person and gene will not be passed down over to the next generations. This approach is currently used in gene therapy. Now for somatic gene therapy, one of the two methods is used.

(i) *Ex vivo* **gene therapy**: This involves removing cells from organs/tissues like skin, haematopoietic system, liver or even tumors and culturing them *ex vivo*, in the laboratory. Now, the therapeutic gene may be introduced by transfecting the cultures/cells by a retroviral vector. The transformed cells are now re-infused or re-implanted back into the patient.

(ii) *In vivo* **gene therapy**: This involves introducing (administering) the gene either locally or systemically into the patients body cells. This method is used for organs like lung, brain or heart, in which case culturing and re-implantation of the cells is not feasible.

So far more than 400 clinical studies (phase I-III) have been approved involving 4000 patients to assess safety of the technique, efficiency of gene transfer methods, sterility and expression of the transferred genes and whether it really provides a 'cure'.

The gene transfer methods are given in Table 2.9. The diseases for which (somatic) gene therapy has been considered are presented in Table 2.10.

Table 2.9 : Methods of gene transfer.

	Viral methods			*Non-viral methods*			
	Retrovirus	*Adenovirus*	*AAV*	**Injection of naked DNA**	**Gene gun**	**Entrapping DNA in liposomes**	**Electroporation**
Genome to be transferred	RNA	DNA	DNA	DNA	DNA, RNA	DNA, RNA	DNA, RNA
Size of genome	<8Kb	<7.5Kb	<5Kb	50Kb<	-	50Kb<	-
Efficiency of method	high	very high	moderate	low	high	low	moderate

AAV: Adeno-assocoiated virus

Table 2.10 : Applications of gene therapy.

Disease	Gene involved and encoding for
Cystic fibrosis	-
Duchenne's muscular dystrophy	-
Familial hypercholesterolemia	LDL-receptor
Atherosclerosis	-
Hemophilia	clotting factor
Cancer of colon, breast, blood, liver, lung	cytokines such as GM-CSF, IFN-γ, IL-2; suicide gene such as Herpes simplex virus-thymidine kinase; tumour suppressor genes such as p53.
Osteoporosis	-
Sickle cell anemia	-
Parkinson's disease	-
Severe Combined Immuno Deficiency (SCID)	cytokine receptor gene; adenosine deaminase
AIDS	-

<table>
<tr><td>3</td><td>

ENZYMES AND ENZYME IMMOBILIZATION

</td></tr>
</table>

CHAPTER CONTENTS AT A GLANCE

3.1   Introduction

Enzymes are substances present in cells in minute amounts and are capable of accelerating biochemical (biological + chemical) reactions associated with life processes. The entire credit, for the successful production of biotechnology based products such as antibiotics, solvents, single cell protein, insulin etc., goes to these catalytic proteins called 'enzymes'. Without these biological catalysts, extremely long and harsh reaction conditions would be required to do the same job or in some cases it even becomes impossible to produce the desired product.

In the natural scheme of things, enzymes have been entrusted with the important task of normal functioning of the organism. Scientists and industrialists have gone a step further and are employing enzymes from diverse sources (see Table 3.1) in one of the two major ways: either medicinally (see Table 3.2) or industrially (see Table 3.3).

Table 3.1 : Sources of enzymes.

I *Organisms producing enzymes as part of their natural physiological mechanism*

(i)	Microbes	: Proteases, lipases, amylases, endonucleases, polymerases etc.
(ii)	Viruses	: RNA-dependent DNA polymerase (reverse transcriptase)
(iii)	Plants	: Pectinase, cellulase, galactomannase etc.
(iv)	Animals	: Trypsin, lysozyme etc.

II *Organisms not normally producing enzyme but induced to produce in cultures*

Bacteria	: Galactosidase, galatomannase etc.

III *Genetically modified organisms producing heterogeneous enzymes*

GM microbes	: Clotting factors, superoxide dismutases, tissue plasminogen activators, galactosidase, glucosidase, recombinant human deoxyribonuclease

Table 3.2 : Enzyme applications: Medicinal and Pharmamaceutical.

I *Therapeutic agents having mechanism of action based on drug-enzyme interaction*

Allopurinol	disulfiram	physostigmine	sulphonamides
asprin	phenelzine	propylthiouracil	theophylline

II *Enzymes having medicinal application*

Asparagenase   collagenase   protease   streptokinase

III *Employed for production of biotech based pharmaceuticals*

The use of REs in genetic engineering or the use of enzymes present in microbial/animal (CHO, insect)/plant cells for production of numerous pharmaceuticals

IV *Bioconversions*

Use of enzyme based transformative capability of microbes to produce high value commercial products such as steroids, prostaglandins, etc.

V *Enzyme dependent diagnostic and analytical tasks*

Performed using/ based on: ELISA, SGOT, SGPT, LDH, glucose oxidase, cholesterol oxidase etc.

VI *Protein engineering*

Involves engineering the genes to design non-natural proteins for improving protein stability, altering antibody properties, altering substrate specificity of an enzyme, or altering pharmacological action.

VII *Rational drug designing*

Involves cloning and expressing the receptor gene (usually on a mammalian cell) and then performing the assay for binding of the receptor to the most appropriate drug molecule.

Table 3.3 : Enzyme applications: Industrial.

Detergents	:	Proteases, lipases
Dairy industry	:	Rennin
Starch processing	:	Amylases (both bacterial and fungal)
Food Industry	:	Not only production of food items but also biopreservation!
Textile processing	:	-
Leather processing	:	Proteases
Biotransformation	:	Various enzymes

3.2   Historical Milestones

In the early decades of 1800, enzymes were referred to as 'ferments' because of their role in brining about fermentation. Though their catalytic role had been postulated but till then enzymes had not been isolated and identified. In fact, Louis Pasteur (1850) was of the opinion that the ferments are inseparable. It was Kuhne (1878) who for the first time coined the term 'enzyme' (which means 'in yeast') to refer to the catalytic substances, the 'ferments', present inside the yeast cells that are responsible for fermentation of sugar into alcohol. Buchner (1897) demonstrated that enzymes are in fact separable. As of today, over 2000 enzymes have been identified. The major milestones in the history of enzymology are presented in Table 3.4.

3.3   Properties and Characteristics of Enzymes

Enzymes are proteins, combined with chemical, non-protein groups. The chemical group are referred to as the prosthetic groups which may be either organic or inorganic. If the prosthetic group is organic it is called as coenzyme and its major integral component is a vitamin (mostly of B complex group). If the prosthetic group is inorganic then it is called as cofactor; the examples of which include metal ions like: Mg^{2+}, Mn^{2+}, Fe^{2+}, Zn^{2+} etc. The protein portion of the enzyme is called as apoenzyme and when united with the non-protein portion, the complete enzyme is called as holoenzyme. The enzymes thus are characterized by having properties akin to the proteins which are given below.

1. Enzymes consist of a protein part, called apoenzyme, which is an inactive, thermolabile, high molecular weight, non-dialyzable molecule.

2. Enzymes consist of a non-protein part, called the prosthetic group, which is an inactive, thermostable, organic or inorganic, low molecular weight, dialyzable molecule.

3. The complete enzyme, called the holoenzyme, is an active molecule.

4. Enzymes are effective catalysts i.e. they speed up biochemical reactions e.g. Carbonic anhydrase, having the highest catalytic activity among the enzymes, catalyzes hydration of 10^5 molecules of CO_2 per second.

5. Enzymes are required in a minute quantity e.g. 1mg trypsin is sufficient to hydrolyze 5lbs of meat, sucrase can hydrolyze 100,000 times its weight of sucrose.

6. Enzymes have specificity for their substrate.

Table 3.4 : Major milestones in enzymology.

Year	Scientist	Achievement / contribution
1837	Jons J. Berzelius	recognized biological catalysis; coined the term 'protein'
1850	Loius Pasteur	studied fermentation of sugar into alcohol by yeast
1878	Friedrich Wilhelm Kuhne	coined the term 'enzyme'; investigated trypsin catalyzed reactions
1894	Emil Fishcer	proposed 'lock & key' hypohesis
1896	Bertrand	studied the requirement of coenzymes/cofactors
1897	Hans Buchner & Eduard Buchner	extracted active form of enzyme from yeast cells
1901	V. C. R. Henri	studied the principle of enzyme-substrate complex; gave the procedure for derivation of kinetic rate laws
1913	Leonor Michaelis & Maud L. Menten	further worked on the theory of enzyme catalysis
1926	J.B. Summner	crystallized urease
1930	John Northrop & Kunitz	crystallized proteolytic enzymes-pepsin and trypsin
1940	Gustav Embden, Otto Meyerhof, Carl Neuberg, Jacob Parnas, Otto Warburg, Gerty Cori & Carl Cori	showed that reactions of lactic acid fermentation in muscle extracts were same as those of alcohol fermentation; elucidated complete glycolytic (Embden-Meyerhof) pathway.
1940	Beadle & Edward Tatum	gave the 'one gene one enzyme concept'
1953	Daniel Koshland	developed the sequential model to explain allosterism
1965	Jacques Monod, Jeffries Wyman, Jean-Pierre Changeux	proposed the concerted (MWC) model to explain allosterism
1965	David Philips	determined three dimensional structure of lysozyme
1970	John C. Gerhart & Arthur Paedee	discovered that ATPase is feed inhibited by CTP

7. Enzymes are very highly sensitive to variations in temperature, pH, concentration of heavy metals, a feature unlike chemical catalysts.

8. Enzymes catalyze the reaction in both the, forward and backward, directions i.e. reversibility of enzyme action.

9. Enzymes are very highly sensitive to variations in temperature, pH, concentration of heavy metals, a feature unlike chemical catalysts.

10. Enzymes are reusable i.e. they remain unconsumed and unaltered after the reaction

11. Enzymes have aqueous solubility.

12. Enzymes have molecular weights ranging from 12,000 to over 1 million.

13. Enzymes are in the colloidal size range.

14. Enzymes are produced by plants, animals, microbes and viruses as endoenzymes or exoenzymes.

15. Enzymes are capable of functioning outside/without the living cells (i.e. exoenzymes).

16. Enzymes have an absolute structural requirement for the catalytic activity: this involves appropriate folding of enzymes to give primary, secondary, tertiary and quaternary structure.

17. Enzymes transform different forms of energy: in photosynthesis, light energy is converted into chemical bond energy i.e. ATP.

3.4   Nomenclature

Enzymes are named based on the different schemes, which are as follows:

1. Original scheme: This was rather non-descriptive of substrate and/or reaction catalyzed e.g. Papain, pepsin, rennin, trypsin, etc.

2. Based on the substrate with '-ase' as suffix: E.g. Amylase, lipase, nuclease, protease, sucrase, etc.

3. Based on the function performed: E.g. Carboxylase, decarboxylase, dehydrogenase, oxidase, transferase etc.

4. Name consisting of two parts: First part stands for the substrate, second part stands for the function performed. E.g. Alcohol dehydrogenase, carbonic anhydrase, monoamine oxidase, thymidilate kinase etc.

5. Enzyme Commission (E.C.) number: This is a rationalized approach to the classification and naming of enzymes. This is the scheme that is currently being followed the world over. It was given by the International Enzyme Commission in 1964. Based on this scheme, enzymes are named in the following manner:

   - All the enzymes are categorized into six major classes depending upon the reaction they catalyze.

   - All the enzymes are given systematic name and a number by the Enzyme Commission.

   - The name and number describe as to what the enzyme does.

- The name consists of two parts: first part specifies the substrate and second part the reaction catalyzed suffixed with - 'ase'.

- The number is a four-digit number.

- The first digit classifies the enzyme into one of the six broad groups:

  i.   Oxido-reductase   :   transfers H atoms or electrons

  ii. Transferase   :   transfers small groups between molecules

  iii. Hydrolase   :   hydrolyzes molecules

  iv. Lyase   :   addition to double bonds

  v.   Isomerase   :   converts between enantiomeric forms

  vi. Ligase   :   forms bonds between C and another atom, using ATP as an energy source

- Each group is subdivided into sub-groups, which is denoted by the second digit.

- Each sub-group is further subdivided into sub-sub-groups, which is denoted by the third digit.

- The last number is specific for the enzyme in that particular sub-sub-group.

- Thus, creatinine kinase is E.C. 2.7.3.2; its systematic name is ATP: creatinine phosphotransferase i.e. an enzyme that transfers (2) a phosphate group (7) from ATP to creatinine (3), and this being the second (2) enzyme in that sub-sub group.

3.5   Catalysis : How?

Enzymes merely accelerate both the forward and backward reactions by precisely the same factor i.e. equilibrium would be achieved earlier. However, they do not bring about any alteration in the equilibrium of a chemical reaction i.e. they will not form more of a product.

Enzymes accelerate biochemical reactions by forming an enzyme-substrate complex which proceeds on to from the product via a transition state having a lower Gibb's free energy of activation i.e. the energy barrier for the formation of the product. The concept is depicted graphically in Figure 3.1.

3.6   Factors Influencing Enzyme Action

Enumerated below are factors that affect the activity of enzymes.

1. Substrate concentration   2. Enzyme concentration

3. Product concentration   4. Temperature

5. pH   6. Time

7. State of oxidation   8. Presence of activators/inhibitors

9. Shear force

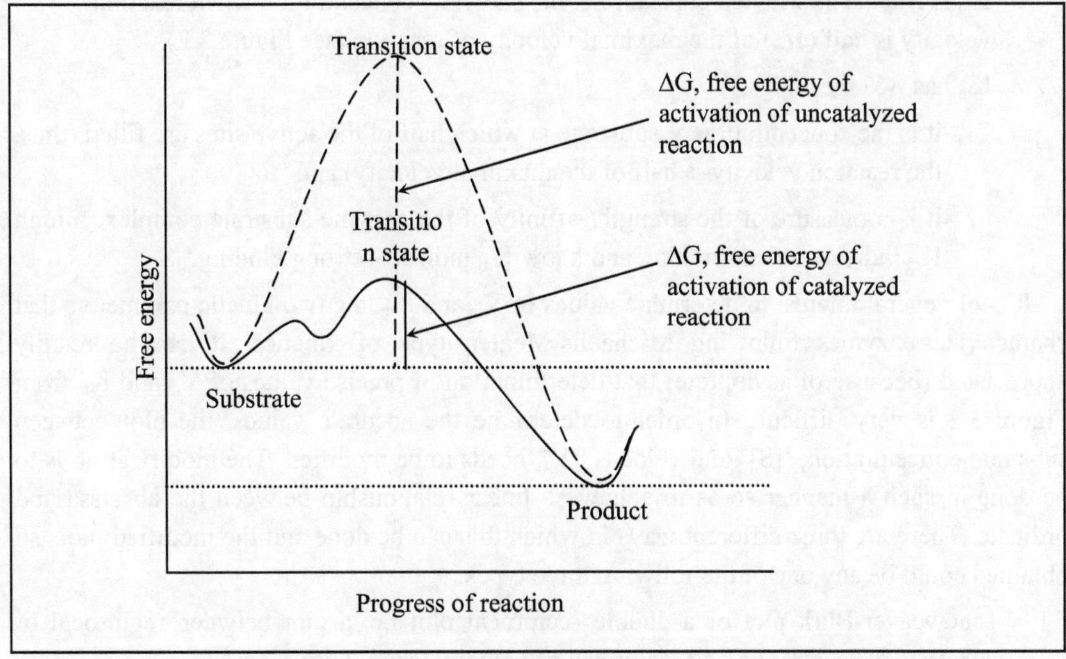

Figure. 3.1 : Acceleration of enzyme catalyzed reaction by lowering of free energy of activation.

3.7   Enzyme Kinetics: Michaelis-Menten or Mixed Order or Saturation Kinetics

In 1902, it was V.C.R. Henri who for the first time studied enzyme kinetics i.e. the rate of enzyme reaction and the physical and chemical conditions affecting it. In 1913, L. Michaelis and M. L. Menten gave a mathematical model to quantify the reaction kinetics based on which, information about mechanism could be obtained and enzyme kinetic parameters could be determined. Their mathematical model is based on single-substrate-enzyme complex, assuming that all the active sites on the enzyme are saturated using high concentration of substrate. The Michaelis-Menten equation describing the velocity of enzyme reaction is:

$$V = \frac{V_m[S]}{K_m + [S]}$$

where,

V is reaction velocity,

V_m is maximum velocity of reaction,

S is substrate concentration and

K_m is Michaelis-Menten constant i.e. the substrate concentration when reaction velocity is half (1/2) of the maximal velocity of reaction (see Figure 3.2).

K_m has two meanings:

1. it is the concentration of substrate at which half of the active sites are filled (thus, the reaction velocity is half of the maximal velocity) and

2. it is a measure of the strength/ affinity of the enzyme-substrate complex: a high K_m indicates weak binding and a low K_m indicates strong binding.

It is of general interest to determine values of V_m and K_m, the two kinetic parameters that characterize enzymes following Michaelis-Menten type of kinetics. It can be readily appreciated (because of asymptote) that determination of precise values of V_m and K_m from Figure 3.2 is very difficult. In order to determine the accurate values, the plot between substrate concentration, '[S]' and velocity 'V', needs to be modified. The modification is to be done in such a manner so as to achieve a linear relationship between the abscissa and ordinate. There are three different ways in which this can be done and the modified plots so obtained could be any one of the follwing three types:

1. Lineweaver-Burk plot or a double reciprocal plot i.e. a plot between reciprocal of substrate concentration and reciprocal of velocity (see Figure 3.3);

2. Eadie-Hofstee plot i.e. a plot between ratio of velocity and substrate concentration and velocity (see Figure 3.4).

3. Hanes-Woolf plot i.e. a plot between substrate concentration and ratio of substrate concentration and velocity (see Figure 3.5).

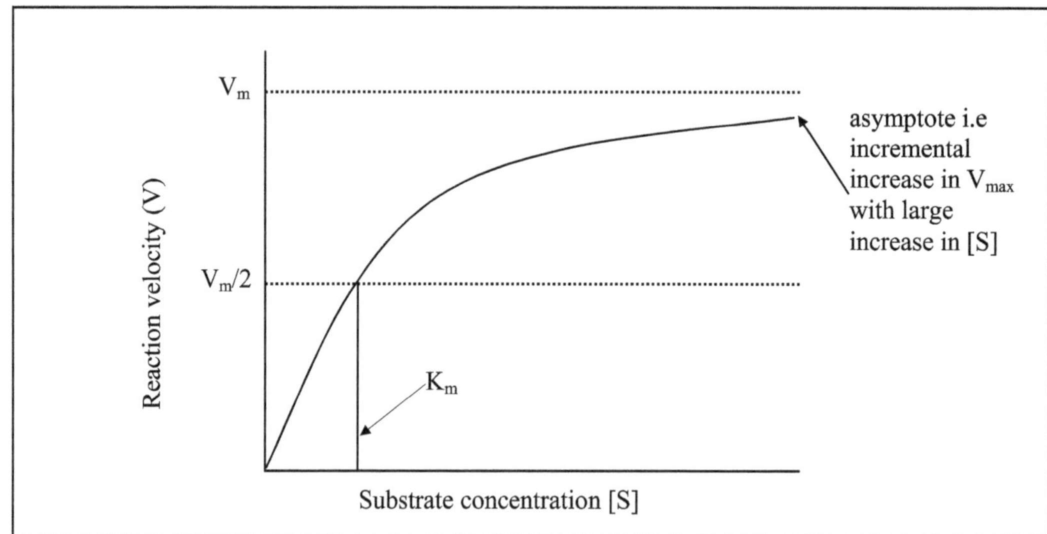

Figure. 3.2 : A plot between substrate concentration and reaction velocity.

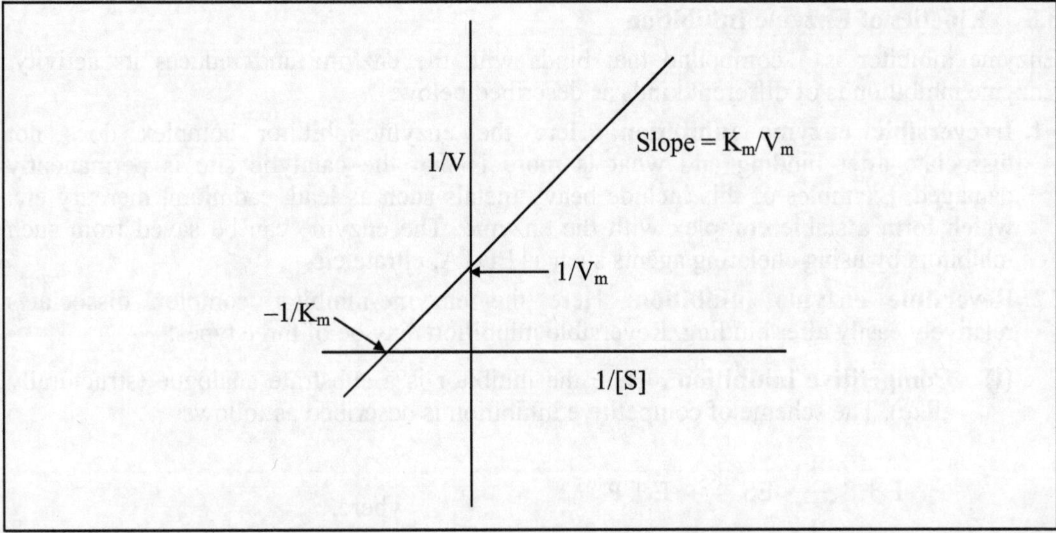

Figure. 3.3 : Double-reciprocal plot or Lineweaver-Burk plot.

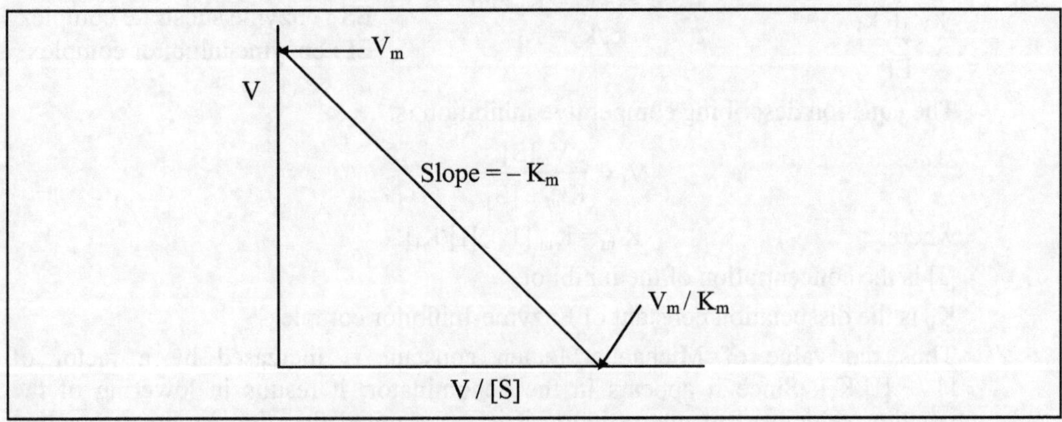

Figure. 3.4 : Eadie-Hofstee plot.

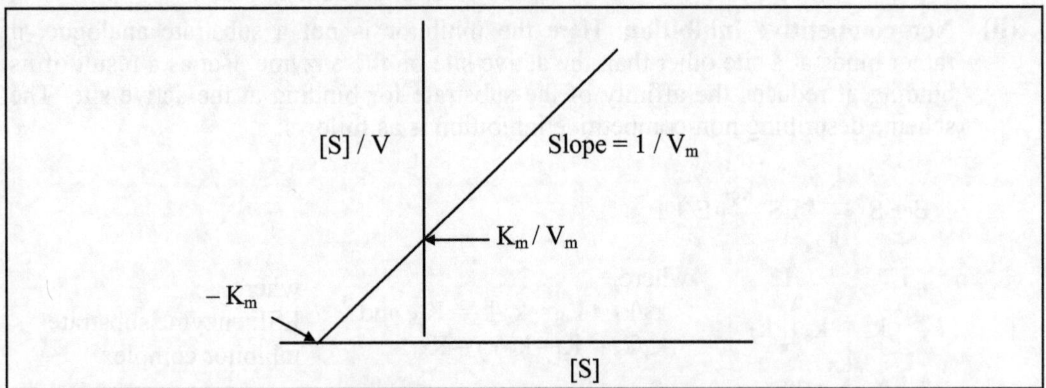

Figure. 3.5 : Hanes-Woolf plot.

3.8   Kinetics of Enzyme Inhibition

Enzyme inhibitor is a compound that binds with the enzyme and reduces its activity. Enzyme inhibition is of different kinds as described below:

1. **Irreversible enzyme inhibition**: Here the enzyme-inhibitor complex does not dissociate after binding and what is more is that the catalytic site is permanently damaged. Examples of this include heavy metals such as lead, cadmium, mercury etc. which form a stable complex with the enzyme. The enzyme can be saved from such inhibitors by using chelating agents such as EDTA, citrate etc.

2. **Reversible enzyme inhibition**: Here the enzyme-inhibitor complex dissociates relatively easily after binding. Reversible inhibition may be of three types:

   (i)   **Competitive inhibition** : Here the inhibitor is a substrate analogue (structurally alike). The scheme of competitive inhibition is described as follows:

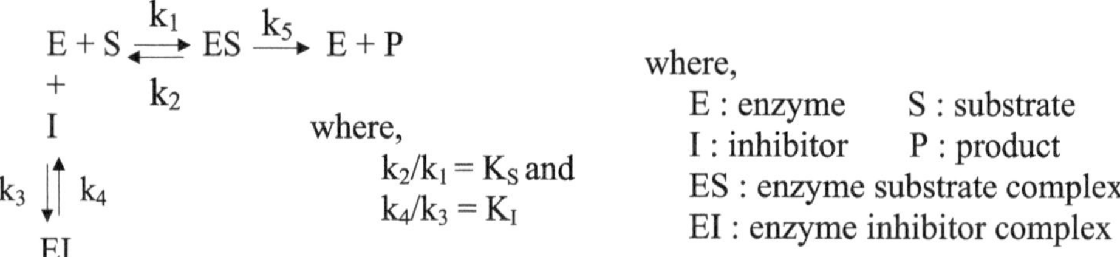

where,

where,

E : enzyme   S : substrate

$k_2/k_1 = K_S$ and

I : inhibitor   P : product

$k_4/k_3 = K_I$

ES : enzyme substrate complex

EI : enzyme inhibitor complex

The equation describing competitive inhibition is:

$$V_I = \frac{V_m[S]}{K'_m + [S]}$$

where, $K'_m = K_m [1 + [I]/K_I]$

[I] is the concentration of the inhibitor

K_I is the dissociation constant of Enzyme-Inhibitor complex.

Thus the value of Michaelis-Menten constant is increased by a factor of $[1 + [I]/K_I]$. Since it appears in the denominator, it results in lowering of the reaction velocity. But interestingly, it is to be noted that the maximum reaction velocity (V_m) remains the unaltered. It is only that a larger amount of the substrate is required to reach the maximum velocity (see Figure 3.6).

   (ii)   **Non-competitive inhibition**: Here the inhibitor is not a substrate analogue, it rather binds at a site other than the active site on the enzyme. But as a result of its binding, it reduces the affinity of the substrate for binding at the active site. The scheme describing non-competitive inhibition is as follows:

$$E + S \underset{k_2}{\overset{k_1}{\rightleftharpoons}} ES \xrightarrow{k_9} E + P$$

where,

$k_2/k_1 = K_S = k_6/k_5 = K_{IS}$ and

where,

$k_4/k_3 = K_I = k_8/k_7 = K_{SI}$

ESI: enzyme substrate inhibitor complex

$$EI + S \underset{k_6}{\overset{k_5}{\rightleftharpoons}} ESI$$

The equation describing non-competitive inhibition is:

$$V_I = \frac{V'_m[S]}{K_m + [S]}$$

where, $V'_m = V_m / [1 + [I]/K_I]$

[I] is the concentration of the inhibitor

K_I is the dissociation constant of Enzyme-(Substrate)-Inhibitor complex.

Therefore, in non-competitive inhibition, the reaction velocity (V_I) and the maximum velocity (V'_m) are both decreased by a factor of $[1 + [I]/K_I]$; while the Michaelis-Menten constant, K_m, remains unaltered (see Figure 3.7).

(iii) **Un-competitive inhibition**: Here the inhibitor binds to the already formed enzyme-substrate complex and as such has no affinity (uncompetitive) for the enzyme. The scheme describing un-competitive inhibition is as follows:

$$E + S \underset{k_2}{\overset{k_1}{\rightleftharpoons}} ES \xrightarrow{k_5} E + P$$

$$+$$
$$I$$

where,

$$k_4 \upharpoonleft\downharpoonright k_3 \qquad k_2/k_1 = K_S \text{ and}$$
$$k_4/k_3 = K_I$$

$$ESI$$

The equation describing uncompetitive inhibition is:

$$V_I = \frac{V'_m[S]}{K'_m + [S]}$$

where, $V'_m = V_m / [1 + [I]/K_I]$ and $K'_m = K_m / [1 + [I]/K_I]$

[I] is the concentration of the inhibitor

K_I is the dissociation constant of Enzyme-Substrate-Inhibitor complex.

Thus in uncompetitive inhibition, the reaction velocity, maximum reaction velocity and the Michaelis-Menten constant are all decreased (see Figure 3.8)

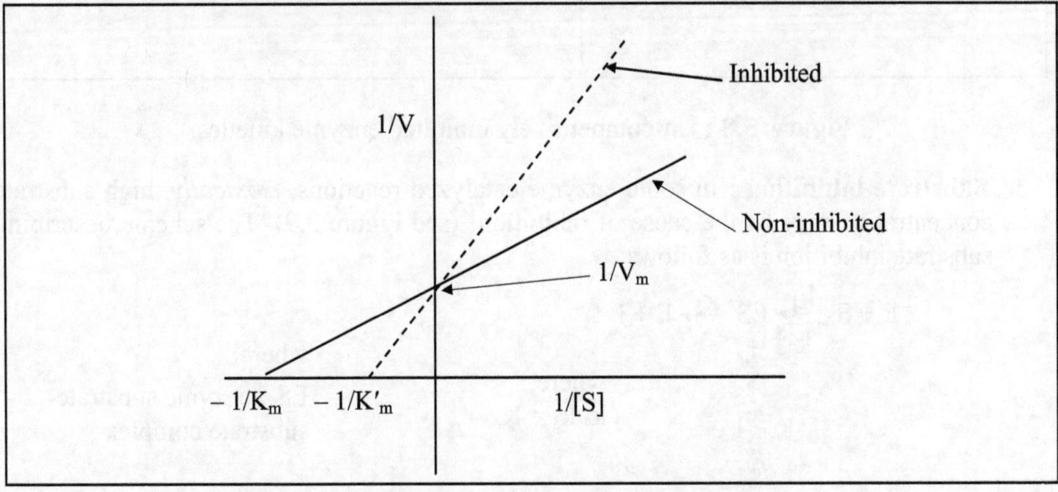

Figure. 3.6 : Competitively inhibited enzyme kinetics.

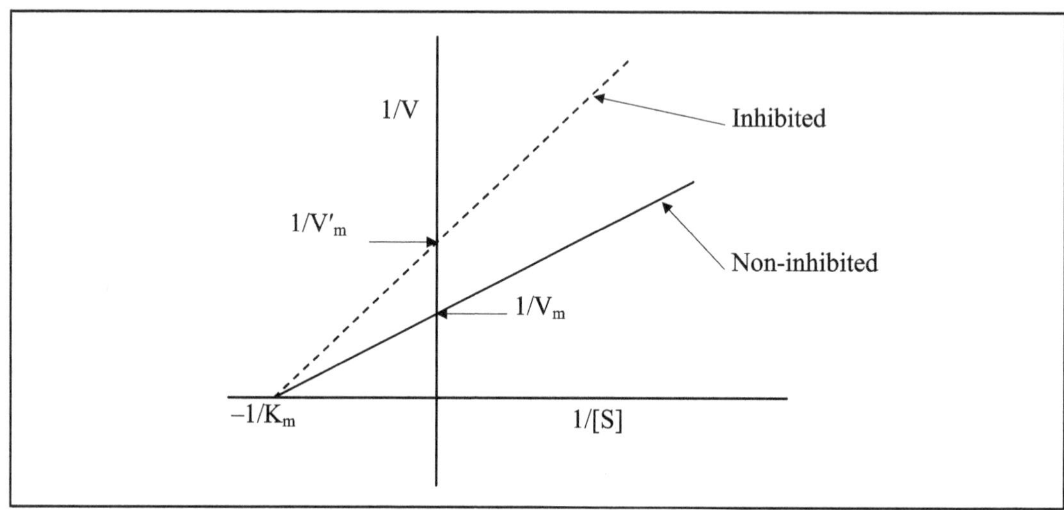

Figure. 3.7 : Non-competitively inhibited enzyme kinetics.

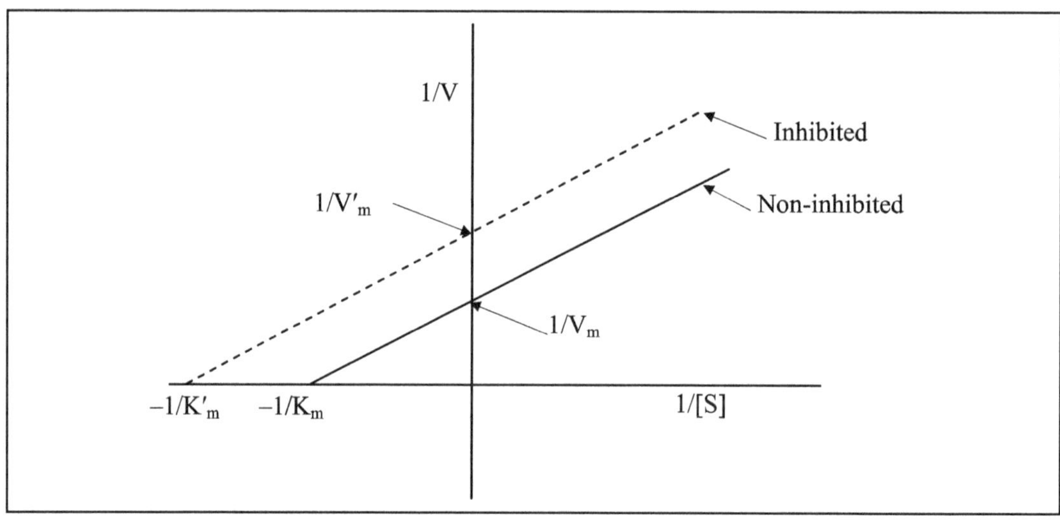

Figure. 3.8 : Un-competitively inhibited enzyme kinetics.

3. **Substrate inhibition** : In some enzyme catalyzed reactions, *ironically*, high substrate concentration may be the cause of inhibition! (see Figure 3.9). The scheme describing substrate inhibition is as follows:

$$E + S \underset{k_2}{\overset{k_1}{\rightleftharpoons}} ES \xrightarrow{k_5} E + P$$

$$+$$

$$S$$

$$k_4 \updownarrow k_3$$

$$ES_2$$

where,

$$k_4/k_3 = K_S$$

where,

ES_2: enzyme substrate-substrate complex

The equation describing substrate inhibition is:

$$V' = \frac{V'_m[S]}{K'_m + [S] + [S]^2/K_I}$$

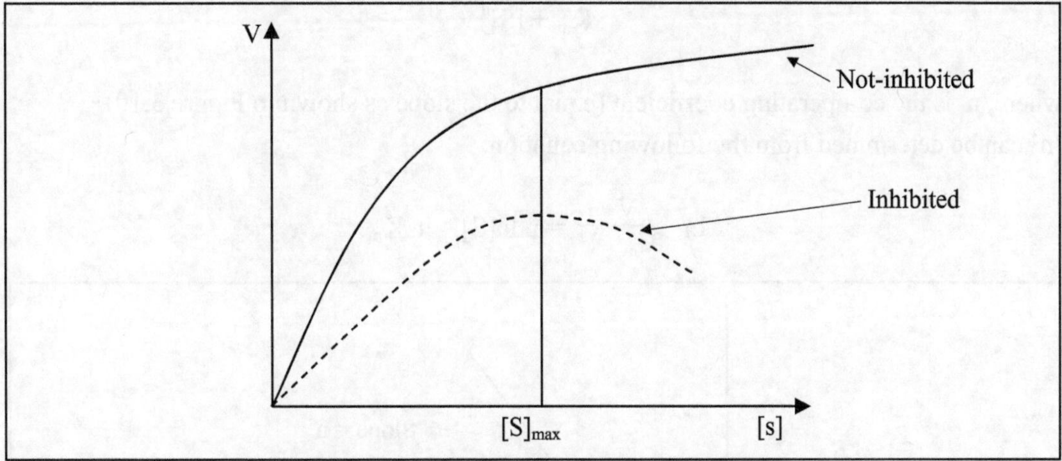

Figure. 3.9 : Substrate – inhibited kinetics.

4. **Suicide inhibition**: Here an enzyme catalyzes the conversion of a substrate into a product that happens to be an inhibitor for that enzyme and it immediately inactivates the enzyme. Thus the catalytic action of the enzyme is said to be suicidal for the enzyme.

   Examples of suicide inhibition include:

   - Inhibition of thymidylate synthase, by flurodeoxy uridylate

   - Inhibition of xanthine oxidase by allopurinol

In both the above cases, flurodeoxy uridylate and allopurinol first act as the substrate for the respective enzymes and then, getting irreversibly and tightly bound to the enzyme, act as inhibitors.

   - Inhibition of monoamine oxidase (MAO) by pargyline-flavin adduct

Here, MAO first catalyzes the formation of pargylline-flavin adduct, once formed, it acts as its inhibitor.

   - Inhibition of glycopeptide transpeptidase by penicillin

The enzyme is responsible for cross-linking of a heteropolymer, peptidoglycan, forming the cell wall of bacteria. On combining with penicillin, it results in its own inhibition, thereby resulting in non-formation of cell wall and ultimately the death of bacteria.

3.9   Allosteric Enzymes

Allosteric enzymes are enzymes that have more than one binding sites i.e. a regulatory site and a catalytic site. The binding of a modulator molecule to the regulatory binding site results in either increased or decreased affinity of substrate for binding with the catalytic site. Generally, the regulatory enzymes (based on feed-back inhibition) are allosteric enzymes. In Figure 3.10, kinetics of allosteric enzymes are depicted and in Figure 3.11, its comparison with the Michaelis-Menten kinetics is depicted. The equation describing reaction velocity of allosteric enzymes is:

$$V = \frac{V_m[S]^n}{K'_m + [S]^n}$$

where, n is the co-operating coefficient (equal to the slope as shown in Figure 3.10).
'n' can be determined from the following equation:

$$\ln \frac{V}{V_m - V} = n.\ln[S] - \ln.K_m$$

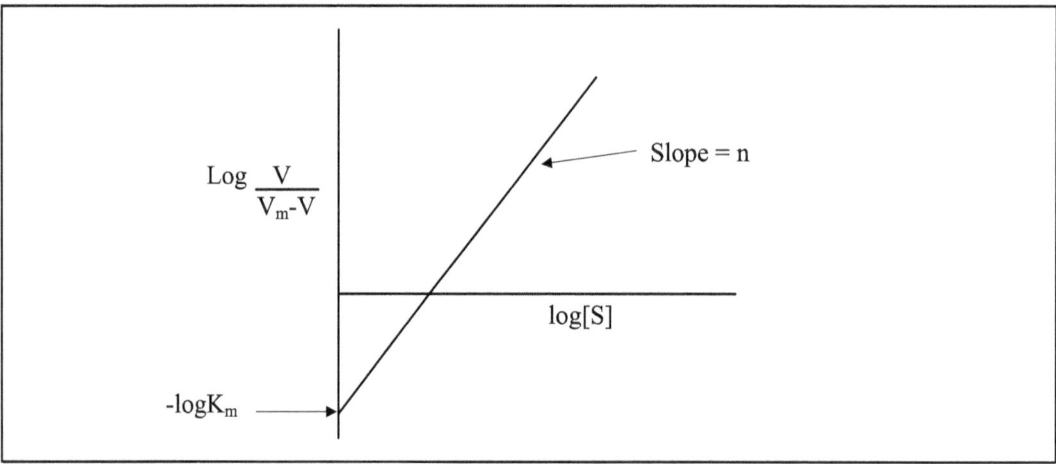

Figure. 3.10 : Kinetics of allosteric enzymes and the co-operativity coefficient .

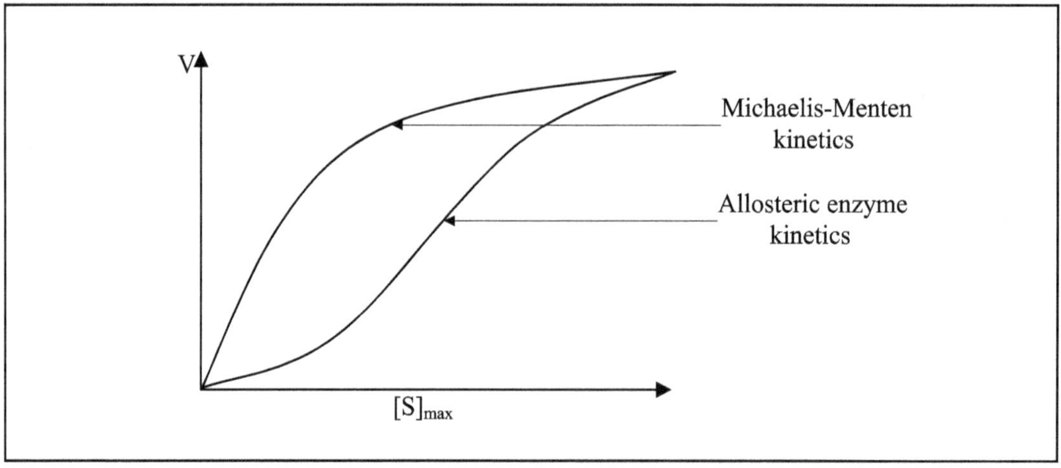

Figure. 3.11: Comparison of Michaelis-Menten and allosteric enzyme kinetics.

3.9.1 Characteristic features

'*Alos*' in Greek means other. Thus, allosteric enzymes are regulatory enzymes having a site other than the catalytic site. The other site is called allosteric site which is specific for the regulator. The regulator, on binding at this site, influences i.e. brings about a conformational change in the catalytic site. The regulator may thus modify/ regulate (i.e. activate or inhibit) the enzyme activity. Further, it is to be noted that allosterism also emphasizes that the regulator need not have any structural resemblance to the substrate. Enumerated below are the characteristics of allosteric enzymes.

1. Allosteric enzymes have both catalytic site and regulatory (allosteric) site.

2. Catalytic site is substrate specific, while allosteric site is modulator specific.

3. Allosteric enzymes have two or more polypeptide chains or subunits.

4. The catalytic and allosteric sites are present on different subunits.

5. Modulator, on binding with the allosteric site, brings about a conformational change in the catalytic site.

6. The enzyme can be stimulated or inhibited by combining with a modulator.

7. Allosteric enzymes are larger and more complex molecules than simple enzymes.

8. They deviate from Michaelis-Menten kinetics. A plot between reaction velocity and the substrate concentration is sigmoidal in case of allosterism rather than hyperbolic as per Michaelis-Menten equation (see Figure 3.11).

9. They experience a feedback inhibition i.e. the end product of multienzyme system acts as inhibitor for the first enzyme.

10. This feedback inhibition is lost on treatment with mercurials, urea, proteolytic enzymes, extremes of pH etc; while the catalytic activity remains intact

11. Allosteric enzymes have enhanced resistance to heat denaturation in comparison to simple enzymes.

12. They undergo reversible inactivation at 0°C.

3.9.2 Mechanism and models

The mechanism of allosterism can be explained based on the following assumptions:

1. Allosteric enzymes are composed of a number of subunits.

2. The subunits assume one of the two states: either 'tense' (T) or 'relaxed' (R) state.

3. The R form has a higher affinity for the substrate, while the T form has lower affinity for the substrate.

4. The increased binding of substrate to the allosteric enzymes (i.e. a high fraction of subunits/ enzymes in R state) can be brought about by either homotropic or heterotropic interactions.

5. A homotropic interaction refers to the binding of substrate molecule to the allosteric enzyme subunit which consequently enhances the binding of subsequent substrate molecules.

6. A heterotropic interaction refers to the binding of a modulator to the allosteric enzyme subunit. The modulator may be positive (stimulatory, activator), stabilizing a high fraction of subunits/ allosteric enzymes in R state thereby resulting in increased binding of substrate molecules. The modulator may be negative (inhibitory, inactivator) stabilizing a high fraction of subunits/ allosteric enzymes in T state thereby resulting in decreased binding of substrate molecules.

7. Thus an allosteric inhibitor shifts the R→T conformational equilibrium towards T, while an allosteric activator shifts it towards R. The concept is illustrated in Figure-3.12.

8. As a result of allosteric interactions with activator or inhibitor, the fraction of enzymes/subunits bound to the substrate molecules is increased or decreased respectively; while the maximum reaction velocity remains unaltered (see Figure-3.13).

The mechanism of regulation of allosteric enzymes is explained on the basis of the following two models:

(i) Monod – Wyman - Changeux (MWC) - Concerted Model

According to this model, the binding of a substrate (homotropic) or modulator (heterotropic) to the subunits of an allosteric enzyme leads to global conformational changes i.e. all the subunits simultaneously shift from T to R state (activation; increased binding with substrate) or from R to T state (inhibition; decreased binding with substrate). The model is schematically depicted in Figure-3.14.

(ii) Koshland-Nemethy-Filmer (KNF) - Sequential Model

According to this model, the transition of state occurs only in the subunits (and not in the whole enzyme composed of a number of subunits) bound with the ligand (i.e. the substrate or modulator). The model is schematically depicted in Figure-3.15. Note that the symmetry is conserved in the concerted model. And it is the concerted model that is widely accepted to explain allosterism.

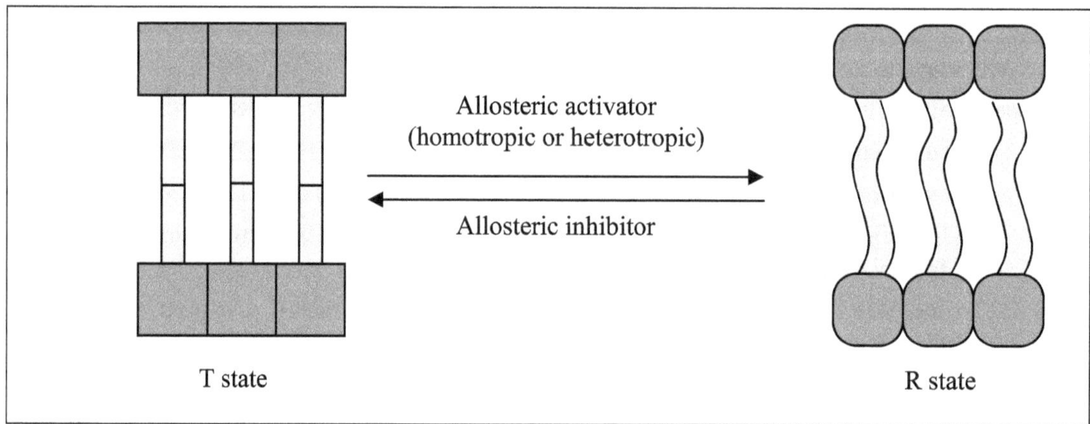

Figure. 3.12 : Shift of equilibrium between T and R forms on combination with modulator.

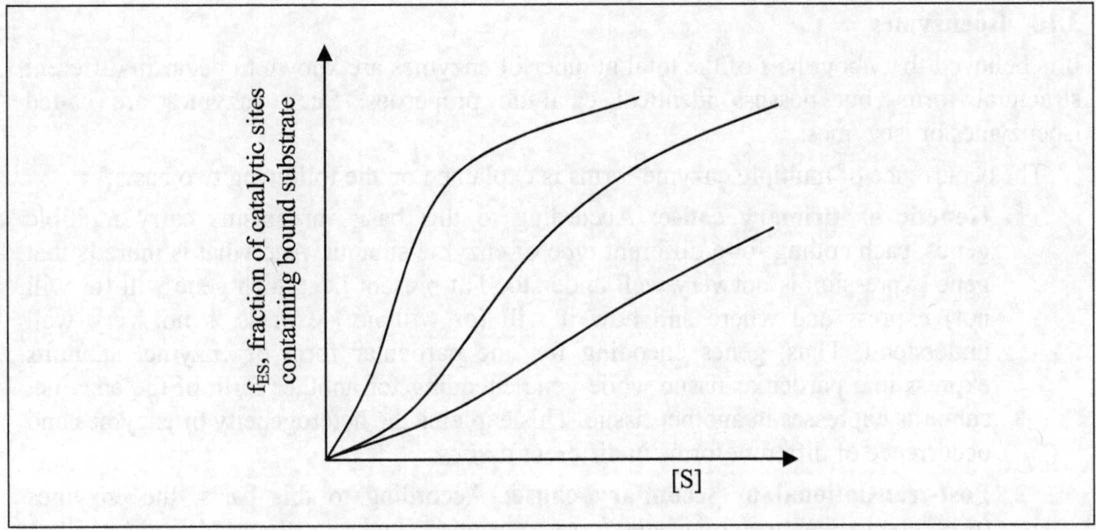

Figure. 3.13 : Dependence of f_{ES} on substrate concentration.

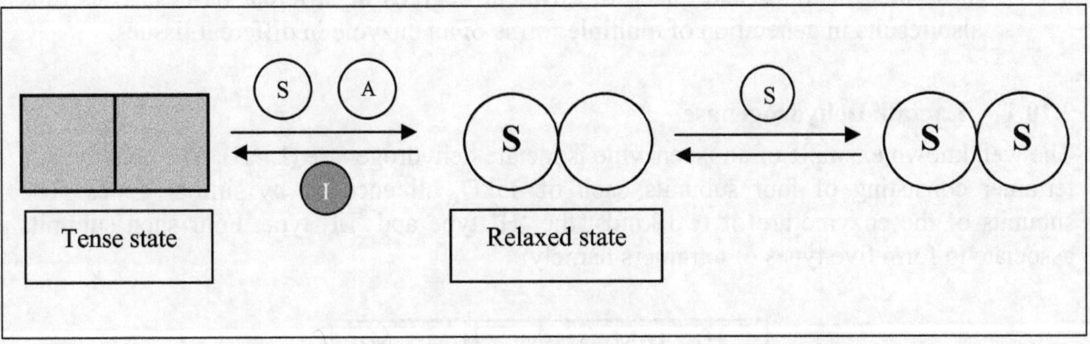

Figure. 3.14 : Monod-Wyman-Changeux - Concerted Model.

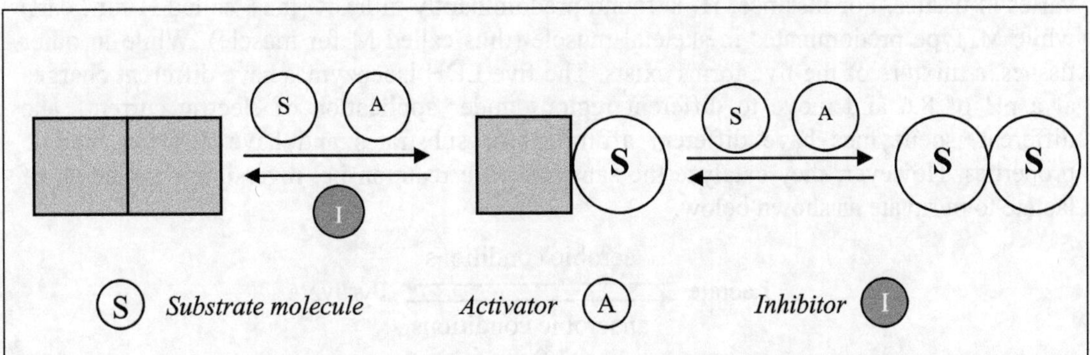

Figure. 3.15 : Koshland-Nemethy-Filmer- Sequential Model.

3.10 Isoenzymes

It is believed that about half of the total number of enzymes are known to occur in different structural forms but possess identical catalytic properties. Such enzymes are called isoenzymes or isozymes.

The occurrence of multiple enzyme forms is explained on the following two basis:

1. **Genetic or primary cause**: According to this basis, organisms carry multiple genes, each coding for a different type of enzyme-subunit. And what is more is that gene expression is not very well understood at present i.e. which gene will (or will not) express and where and how it will (or will not) express is not very well understood. Thus, genes encoding for one particular form of enzyme/ subunits express in a particular tissue while gene encoding for another form of the enzyme/ subunits expresses in another tissue. This explains the heterogeneity of enzymes and occurrence of different forms in different tissues.

2. **Post-translational or secondary cause**: According to this basis, the enzymes (proteins) having been translated (post-translation) undergo structural modifications. The modifications include addition of carbohydrate, limited proteolysis, covalent modification of amino acid side chains, folding etc. Here again, it is known that these modifications take place to different degrees in different tissues. Thus, this also results in generation of multiple forms of an enzyme in different tissues.

3.10.1   Lactate Dehydrogenase

The well known example of an isoenzyme is lactate dehydrogenase (LDH). The enzyme is a tetramer consisting of four subunits each of 35kD, all encoded by similar genes. The subunits of the enzyme are of two kinds: the 'H' type and 'M' type. Four such subunits associate to form five types of tetramers namely:

$$\boxed{H_4, \quad H_3M_1, \quad H_2M_2, \quad H_1M_3, \quad M_4}$$

These five species are isoenzymes of LDH. The distribution of these forms in an individual varies in tissues. For instance, H_4 is found predominantly in heart (thus, called H for heart) while M_4 type predominates in skeletal muscle (thus called M for muscle). While in other tissues, a mixture of the five forms exists. The five LDH isoenzymes have different charges at a pH of 8.6 and move to different regions under application of electric current. The different isoenzymes have different affinities for substrates and have different kinetic properties. However, they catalyze the same specific reaction i.e. reversible conversion of lactate to pyruvate as shown below.

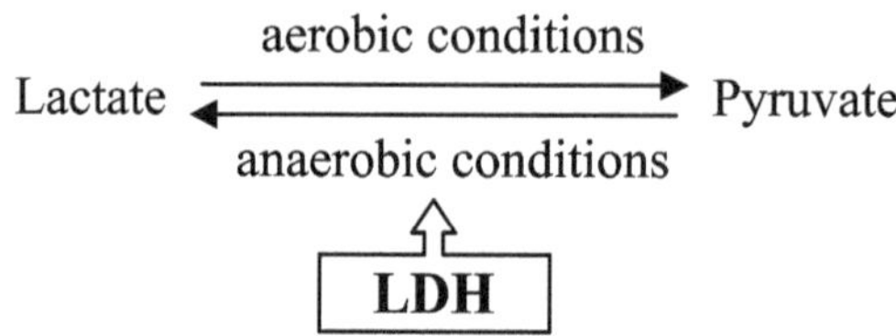

The H_4 LDH isoenzyme for instance has a higher affinity for lactate than the M_4 LDH isoenzymes. Further, H_4 is mainly responsible for catalyzing the conversion of lactate to pyruvate under aerobic conditions. It is inhibited by high levels of pyruvate. On the other hand, M_4 catalyzes the conversion of pyruvate to lactate under anaerobic conditions.

The determination of levels of LDH isoenzyme in different tissues is helpful in diagnosis of muscle, heart or liver diseases. Likewise, creatinine kinase (CK), malate dehydrogenase (MDH), leucine aminopeptidase etc. all occur in multiple forms as isoenzymes. The determination of levels of these enzymes helps in the diagnosis of diseases like hepatic disease, pernicious anemia, muscular dystrophy, myocardial infarction, cancer etc.

3.11 Antibiotic Inactivating Enzymes

These are enzymes produced by the microbes that are responsible for inactivating the antibiotics and thereby resulting in a therapeutic failure. Some such enzymes are considered below.

1. **Penicillanase:** The enzyme is responsible for inactivating penicillins and cephalosporins. The penicillanases are of two types:
   - Penicillin acylase or amidase: Amongst the various genera, these are notably produced by *Penicillium*, *Aspergillus* and *Mucor*. These act on the amide bond via which the acyl group is attached to the basic nucleus i.e. 6-amino penicillanic acid (6-APA).
   - β-lactamase: These are notably produced by *Streptococcus* sps. These attack the β-lactum ring of the basic nucleus 6-APA and rupture it giving penicilloic acid.

2. **Chloramphenicol acetyltransferase:** This enzyme is responsible for the inactivation of chloramphenicol. The enzyme catalyzes the acetylation of the antibiotic in a two stage process. The resulting product, 1,3-diacetylchloramphenicol, is inactive.

3. **Aminoglycoside inactivating enzymes:** Inactivation of aminoglycoside antibiotics (such as streptomycin, gentamicin, kanamycin, neomycin etc.) is brought about by one of the following enzyme catalyzed reactions:
   - Gentamicin acetyl transferase- N-acetylation of amino groups in gentamicin
   - Kanamycin acetyl transferase- N-acetylation of amino groups in kanamycin A, B, neomycin, gentamicin
   - Gentamicin adenyl transferase-adenylation of hydroxyl groups in gentamicin, kanamycin
   - Streptomycin adenyl transferase-adenylation of hydroxyl groups in streptomycin, spectinomycin
   - Kanamycin phosphotransferase-phosphorylation of hydroxyl groups in kanamycin, neomycin, gentamicin A
   - Lividomycin phosphotransferase-phosphorylation of hydroxyl groups in neomycin, paramomycin

- Streptomycin phosphotransferase-phosphorylation of hydroxyl groups in streptomycin

3.12 Study of some Enzymes

Presented in this section, is a brief study on some of the common enzymes.

1. **Hyaluronidase:** This enzyme catalyzes the breakdown of hyaluronic acid. Hyaluronic acid is a mucopolysaccharide and is an essential intracellular tissue cementing material that binds together the parenchymal cells of organs/tissues. Hyaluronic acid though present in all organs, occurs predominantly in connective tissue and blood vessels. Amongst the various genera, the enzyme is notably produced by *Clostridium perfringenes*, the gas-gangrene causing microbe. By virtue of enzyme's hydrolyzing property, the microbes are able to further penetrate into the host tissue from the site of infection. Hence, the enzyme is commonly known as 'spreading factor'.

   In mammals, hyaluronidase is present in the highest concentration in testes. Hence, commercially hyaluronidase is prepared from bovine testes. In therapy, its role of hydrolyzing hyaluronic acid is used to advantage in that it promotes the 'spreading' of drug particles or solution over a large area in tissue. As a result, enhanced absorption of the hypodermically administered drug is achieved in tissue spaces, transudates and edema. However, caution has to be exercized by not administering hyaluronidase in the infected tissue; and not in combination with local anaesthetics.

2. **Streptokinase:** This enzyme is responsible for the conversion of plasminogen to plasmin. Plasmin in turn dissolves the fibrin of blood clots. The enzyme is notably produced by β-haemolytic *Streptococci*. It aids the spread of infection by dissolving the formed clot at the damaged site of infection.

   In therapy, its action of dissolving clots is used to advantage as thrombolytic agent for dissolving coronary thrombi and thereby restoring coronary perfusion. For successful life-saving fibrinolytic action, it has to be administered within 4-6 hours of formation of clot (which is manifested as myocardial attack).

3. **Amylases:** A group of enzymes that hydrolyze complex sugars (such as starch, sucrose etc.) into simple sugars (such as glucose, fructose etc.) are termed glycosidases. One of the major groups of these enzymes is amylases which break down long chain starch and similar polymers into shorter segments and glucose. Amylases are notably extracted from barley, beans, potato and produced by microbes: bacteria and fungi both. The amylases produced by microbes are used in industries such as textile, paper, food, brewing and detergent (for removing food spots in dry-cleaning in conjunction with proteases).

The amylases from the two microbial sources are dissimilar and a comparison of the two is given in Table 3.5.

Table 3.5 : A comparison of microbial amylases.

Basis of comparison	Bacterial amylases	Fungal amylases
Produced by	*Bacillus subtilis* *Bacillus mesentericus*	*Aspergillus niger* *Aspergillus oryzae*
Stability and activity	up to 90°C	deactivated at 55°C
Maximum activity at pH	of 6.5 - 7.0	of 4.0 - 5.0
Type of activity	liquefying enzyme i.e. it depolymerizes starch and produces rapid fall in viscosity of dispersion	saccharifying enzyme i.e. it produces reducing sugars such as maltose
End product	dextrans	maltose
Method of production	sub-merged culture	solid-substrate culture (Tray process)

4. Protease: These are the enzymes responsible for the breakdown of proteins and include proteinases and peptidases. Proteinases are exoenzymes being excreted out of the microbial cell into the fermentation medium; while peptidases are liberated only on autolysis of microbial cells. Among the two, it is the proteinases that are of commercial interest. The proteinases are classified into four groups as given in Table 3.6.

Table 3.6 : Classification of proteinases.

Class	Example	Source
Serine (alkaline) proteinases	trypsin, chymotrypsin subtilisin	animal pancreas *Bacillus sublius licheniformis*, *Bacillus subtilis*, *Bacillus amyloliquefaciens*
Thiol proteinases	papain bromelain	papaya pineapple
Carboxyl (acid) proteinases	fungal proteases rennin pepsin	*Aspergillus* spp., *Mucor* spp. calf stomach pig stomach
Metallo-(neutral) proteinases	thermolysins fungal proteases	*Bacillus thermoproteolyticus*, *Bacillus subtilis*, *Bacillus amyloliquefaciens* *Aspergillus* spp.

Proteases tend to be highly specific for the substrates. They are employed:

- in detergents (laundering),
- for bating of hides in leather industry (tanneries),

- for tenderizing meat,
- for food processing,
- as food supplements,
- for synthesis of peptides by organic phase catalysis,
- as therapeutic agents for breakdown of blood clots (as thrombolytic agents) and
- as drug targets (e.g. Angiotensin converting enzyme, HIV proteases).

Some of the therapeutically important proteases are presented in Table 3.7.

Table 3.7 : Some proteases of therapeutic importance.

Enzyme	Source	Application/ Indications
Bromelain	*Anaanas comosus*	inflammation, edema
Chymotrysin	Bovine pancreas	inflammation, edema, upper RTI, ophthalmology,
Pancreatin	Animal pancreas	pancreatitis
Papain	*Carica papaya*	dyspepsia, gastritis
Plasmin	Plasminogen	as anticoagulant in thrombolytic disorders
Streptokinase	*Streptococci* sps.	thromboembolic diseases
Trypsin	Animal pancreas	cleaning necrotic tissue
Tissue plasminogen activator	rDNA technology	thromboembolic diseases
Urokinase	human urine	thromboembolic diseases

The major bulk of the proteases are produced using microbes: fungi or bacteria. The fungal proteases have a wider pH activity range than bacterial or even animal protease. As a result of this, fungal proteases are more often employed commercially.

3.13   Enzyme Immobilization

Enzyme immobilization is defined as the restriction over the movement or mobility of the enzyme in a fixed space. Synonymously, it is known as enzyme restriction or homing.

It is always sensible to immobilize enzymes before use rather than use enzymes as such. The advantageous reasons for using immobilized enzymes are threefold:

1. **Reusability:** The immobilized enzyme can be easily recovered from the reaction mixture and can be reused, since the enzyme is not consumed in the reaction. Also, enzymes are expensive and cannot be allowed to go down the drain.

2. **Pure product:** The product of enzyme action, as recovered, is pure (free of enzyme); thereby eliminating the need for expensive purification process.

3. **Suitable environment and increased efficiency:** The entrapped enzyme is in an environment very much like its natural environment inside the cell and therefore performs better. Generally, enzymes have better catalytic efficiency when they are attached to a surface.

3.13.1   Materials and Methods

There are two major methods of immobilization of enzymes as shown in Figure 3.16. The support materials employed for enzyme immobilization are presented in Table 3.8.

I. Immobilization on surface of a support.

1. Adsorption
    (physical, involving van der Waals forces)

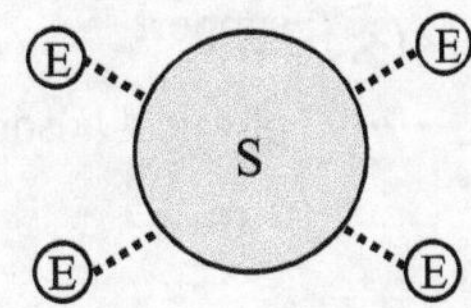

2. Covalently bound

3. Abzymes:binding with monoclonal antibody-carrier complex

4. Co-polymerization

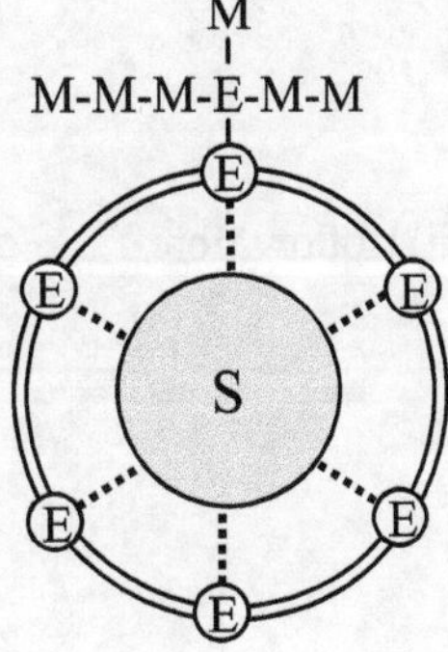

5. Using multifunctional reagents
    i)   Physical adsorption followed by cross linking

    ii) Introduction of functional groups on the support
         followed by covalent reaction with enzyme

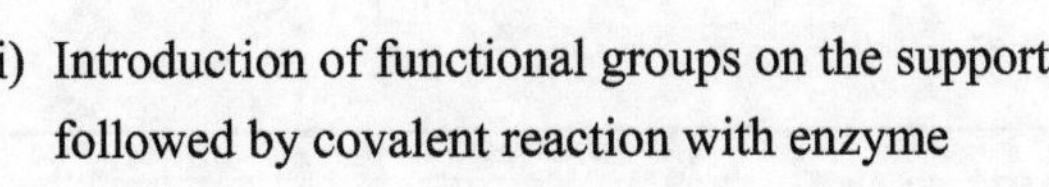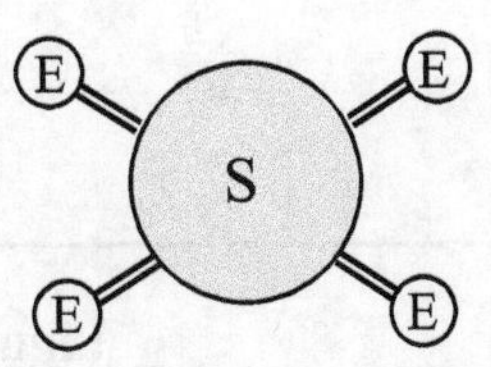

Contd...

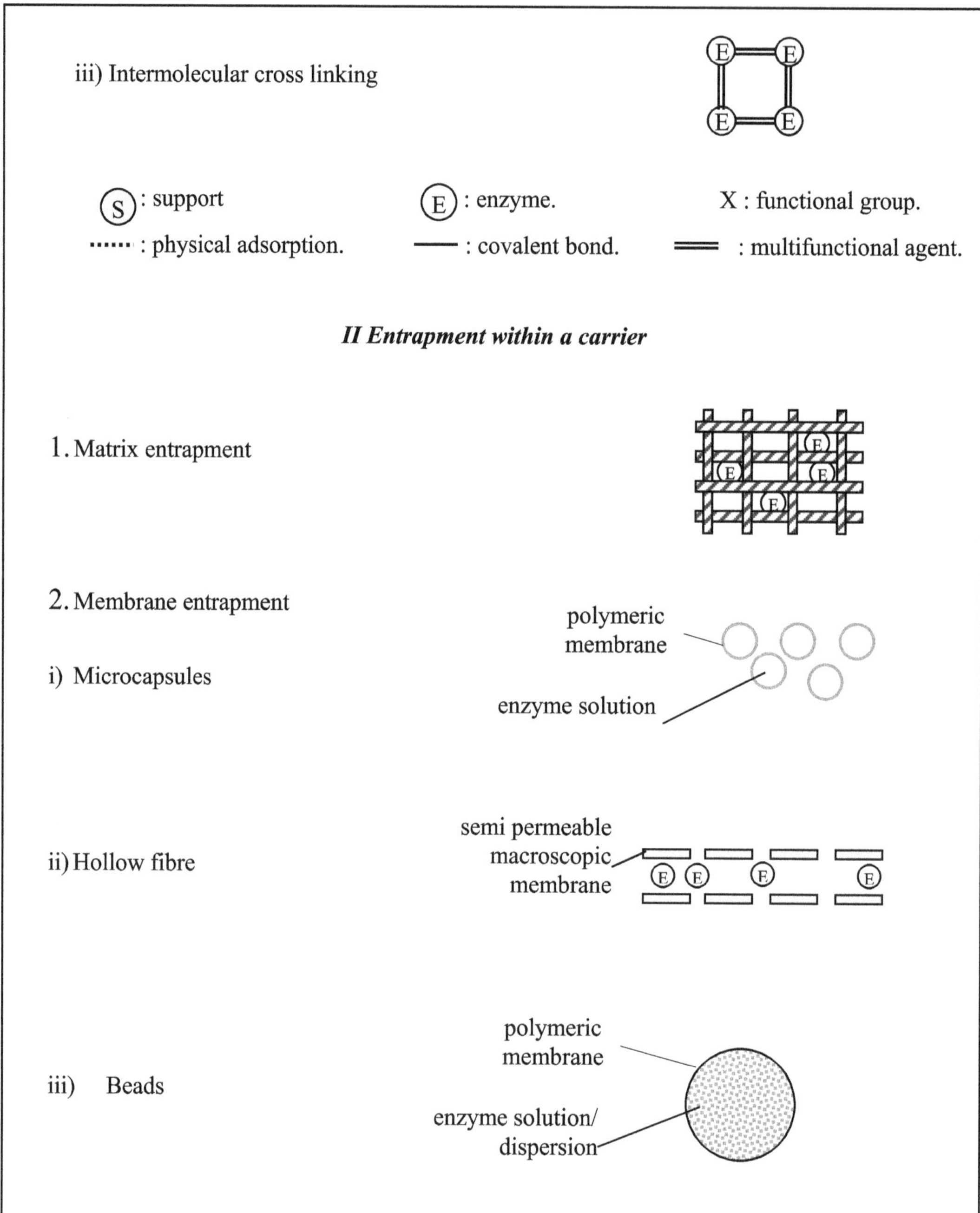

Figure. 3.16 : Methods of enzyme immobilization.

Table 3.8 : Support materials for enzyme immobilization.

Class	Examples
1. Adsorbents	
i) Inorganic	Alumina, silica, glass, ceramics, diatomaceous earth, clay, bentonite, calcium carbonate
ii) Organic	CMC, DEAE cellulose, starch, charcoal, collagen
iii) Ion-exchange resins (anionic and cationic)	Amberlite, Sephadex, Dowex
2. Support materials with functional groups for covalent binding	
i) Supports with-NH_2 group	Natural: Agarose, cellulose, dextran, glass, starch
ii) Supports with-COOH group	Synthetic: Polymers based on acrylamide, styrene
iii) Supports containing anhydride	methacrylic acid, maelic-anhydride,
iv) Supports with-phenyl group	
v) Supports with-hydroxyl group	
vi) Supports with-sulphydryl group	
vii) Supports with-imidazole group	
3. Abzymes	
i) Monoclonal antibodies covalently bound to carriers	Eupergit C, cellulose, sepharose, protein A
4. Multifunctional agents for cross-linking	
i) Bifunctional	Glutraldehyde, bis-diazobenzidine, 2,2-disulphonic acid
5. Matrix materials	
i) Polymeric materials	Calcium alginate, agar, kappa-carragenan, polyacrylamide, collagen
ii) Solid material	Charcoal, ceramic, diatomaceous earth
6. Membrane forming materials	
	Nylon, cellulose, polysulphone, polyacrylate

3.13.2   Notes on materials and methods

1. *Adsorption and adsorbents*

- A simple method
- Deactivation of enzyme is unlikely, since active site is unaffected
- Reversible process i.e. desorption is common

- Weak bonding involving Vander Waal's forces, hence easily desorbed upon little shear
- Percentage of enzyme immobilization is low
- Cross linking by multifunctional agents stabilizes adsorption
- Physical or chemical pretreatment of the adsorbent surface is needed prior to immobilization

2. *Covalent binding and materials with functional groups*

- Covalent bond formation is involved between the functional groups such as amino, carboxyl, hydroxyl, sulphydryl etc., present on the enzymes and the support materials (see Table 3.8)
- Prior to binding, activation of the functional groups on the support material is needed. Activation is brought about by using chemical reagents such as cyanogen bromide, carbodiimide and glutraldehyde (see Table 3.9).
- The chance of loss of enzymatic activity is high, if the active site of the enzyme becomes involved in the chemical reaction.

3. *Covalent binding and antibody carrier complex*

- This method has the advantage of specifically binding with the desired enzyme
- The binding is such that activity of the enzyme is retained

4. *Cross linking and multifunctional agents*

- Accompanied with significant changes in active site of enzymes
- Note that glutraldehyde, a cross linking agent, is also a protein denaturant

5. *Matrix and membrane entrapment and polymeric materials*

- A highly cross linked network of polymers, i.e. gel formation, that entraps the enzyme
- There is no chemical modification of enzyme in this method
- Hence, intrinsic enzyme properties are not altered
- Chances of enzyme deactivation are high during gel (matrix) formation
- Enzyme leakage out of the matrix/membrane is a problem
- Diffusional limitations are highest in beads> followed by microcapsules >and least in matrices
- There is a possible disadvantage of reduced activity and stability due to unfavorable micro-environmental conditions

3.14 Applications of Enzyme Immobilization

The applications of enzyme immobilization are presented in Table 3.10.

Table 3.9 : Chemical reagents used for activation of support materials and covalent binding of enzymes.

Supports with — OH

1. Using cyanogens bromide

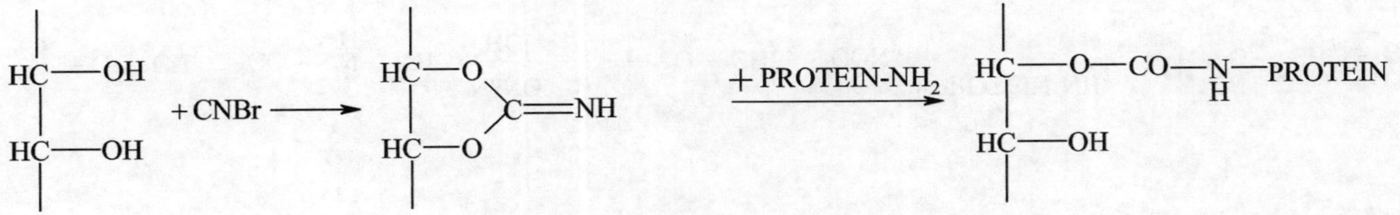

$$HC-OH,\ HC-OH\ +\ CNBr \longrightarrow\ HC-O,\ HC-O,\ C=NH\ \xrightarrow{+\ PROTEIN-NH_2}\ HC-O-CO-NH-PROTEIN,\ HC-OH$$

2. Using *S*-triazine derivatives

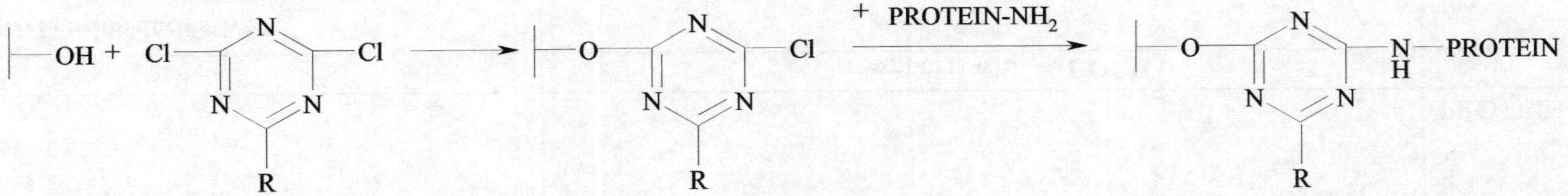

$$OH\ +\ Cl-\text{(triazine)}-Cl \longrightarrow O-\text{(triazine)}-Cl\ \xrightarrow{+\ PROTEIN-NH_2}\ O-\text{(triazine)}-NH-PROTEIN$$

(triazine ring bearing R substituent)

Supports with — NH_2

1. By diazotization

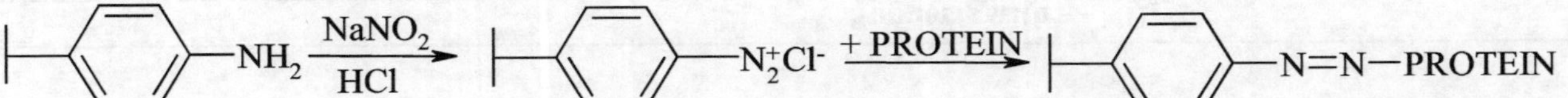

$$\text{(support)}-\text{C}_6\text{H}_4-NH_2\ \xrightarrow[\text{HCl}]{NaNO_2}\ \text{(support)}-\text{C}_6\text{H}_4-N_2^+Cl^-\ \xrightarrow{+\ PROTEIN}\ \text{(support)}-\text{C}_6\text{H}_4-N=N-PROTEIN$$

Table 3.9 Contd…

Supports with — NH_2

2. Using glutraldehyde

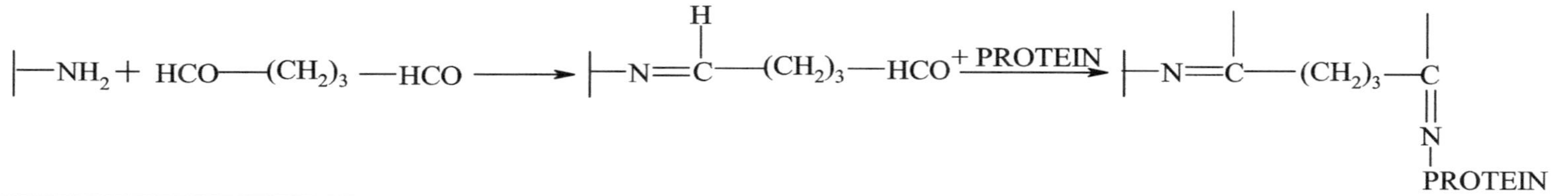

$$\text{—}NH_2 + HCO\text{—}(CH_2)_3\text{—}HCO \longrightarrow \text{—}N{=}\overset{H}{C}\text{—}(CH_2)_3\text{—}HCO \xrightarrow{+\ PROTEIN} \text{—}N{=}C\text{—}(CH_2)_3\text{—}C{=}N\text{—PROTEIN}$$

Supports with — COOH

1. Via azide derivative

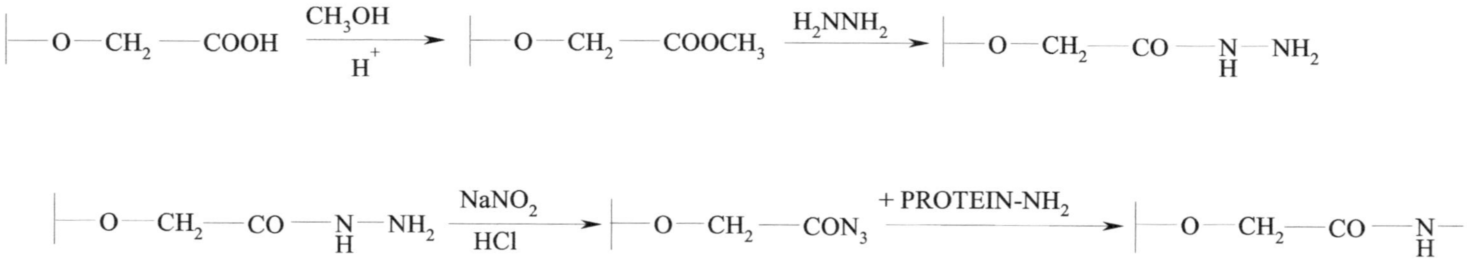

$$\text{—}O\text{—}CH_2\text{—}COOH \xrightarrow[H^+]{CH_3OH} \text{—}O\text{—}CH_2\text{—}COOCH_3 \xrightarrow{H_2NNH_2} \text{—}O\text{—}CH_2\text{—}CO\text{—}\underset{H}{N}\text{—}NH_2$$

$$\text{—}O\text{—}CH_2\text{—}CO\text{—}\underset{H}{N}\text{—}NH_2 \xrightarrow[HCl]{NaNO_2} \text{—}O\text{—}CH_2\text{—}CON_3 \xrightarrow{+\ PROTEIN\text{-}NH_2} \text{—}O\text{—}CH_2\text{—}CO\text{—}\underset{H}{N}\text{—}$$

2. Using carbodiimide

$$\text{—}COOH + \underset{N-R_2}{\overset{N-R_1}{C}} \longrightarrow \text{—}\underset{N-R_2}{\overset{O}{\underset{\|}{C}}}\text{—}O\text{—}\overset{HN-R_1}{C} \xrightarrow{+\ PROTEIN\text{-}NH_2} \text{—}\overset{O}{C}\text{—}\underset{H}{N}\text{—PROTEIN}$$

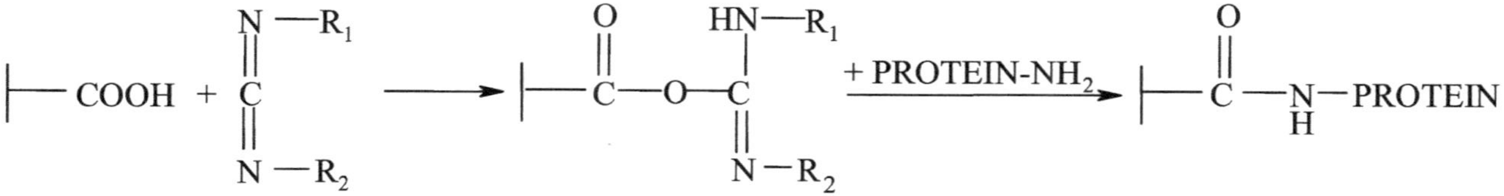

Table 3.10 : Applications of enzyme immobilization.

Enzyme	Method of immobilization	Application
Industrial		
Glucose isomerase	embedment in collagen matrix concentration	isomerization of glucose to fructose in production of high fructose syrup which is sweeter
Aminoacylase	-	resolution of racemic (DL) mixture of amino acids to give L-amino acids which have nutritional/ physiological value
Analytical		
Glucose oxidase	entrapment in polyacrylamide gel and layered over O_2 electrode	determination of glucose concentration (glucose electrode)
Urease	layered on Platinum/ O_2 electrode	determination of urea and uric acid concentration (urea electrode)
Lactate dehydrogenase	layered on Platinum/ ferricyanide electrode	determination of lactate concentration
AMP deaminase	layered on NH^+ electrode	assay of adenosine monophosphate
Alcohol oxidase	layered on Platinum/ O_2 electrode	assay of alcohols
Alkaline phosphatase	layered on Platinum/ O_2 electrode	assay of glucose-6-phosphate
Penicillinase	layered on pH electrode	assay of penicillin
Glucose-6-phosphate dehydrogenase	-	autoanalysis of glucose
Horseradish peroxidase	-	ELISA, the immobilized enzyme is conjugated with an antibody

Table 3.10 Contd...

Therapeutic		
Fibrinolysin/ urokinase Streptokinase	adsorbed on Sephadex	breakdown of thromboemboli
α-amylase, catalase, trypsin chymotrypsin, ribonucleas	adsorbed on dextran	have slower clearance from blood
Phenylalanine ammonia lyase	encapsulation in microcapsules, fibres or gel	reduction of blood levels of phenylalanine
Heparinase	bound to Sheparose	used in blood purifying extracorporeal shunt
Carboxy peptidase	covalently bound to dextran/PEG	-
Arginase, urease	covalently bound to dextran/PEG	-
Lysosomal hydrolase	-	therapy of Tay - Sachs disease
β-D-N-acetyl hexoaminidase	-	
11- β- hydroxylase with 11-deoxycortisol + Δ^1 dehydrogenase	all encapsulated together	treatment of arthritis
Urease + adsorbent resin or charcoal	all microencapsulated together	compact artificial kidney
Urease + urea (yields ammonium bicarbonate & ammonium hydroxide)	entrapped in copolymer of methyl vinyl ether + maleic anhydride	as novel drug delivery system based on pH (alkaline) dependent erosion of copolymers
Glucose oxidase + glucose insulin	entrapped in poly(ortho esters) (yields gluconic acid)	as novel drug delivery system for insulin based on pH (acidic) dependent erosion of copolymers
β- galactosidase, invertase, urease, arginase, β-fructosidase, asparaginase	entrapped in erythrocytes (Gaucher's disease)	enzyme replacement therapy and lysosomal storage diseases

4 # FERMENTATION

CHAPTER CONTENTS AT A GLANCE

4.1   Introduction

Fermentation is a discipline of traditional biotechnology. The word fermentation comes from the Latin term *'ferver'*, which means 'to boil'. The word has been appropriately coined, but of course in its traditional sense, to describe the catabolic action of yeast on sugars to produce alcohol accompanied with evolution of carbon dioxide bubbles, resembling the boiling appearance. Microbiologically, rather strictly and traditionally, fermentation involves the use of metabolic capacity (i.e. enzymes) of microorganisms on organic substrates under anaerobic conditions to produce an economically useful metabolite.

However, in modern terms, fermentation applies to the growth of microorganisms (and not plant and animal cells) in liquid media under either condition (aerobic or anaerobic) to produce economically useful products. Alternatively, if defined in its biochemical sense, fermentation means a process by which microorganisms generate energy from organic compounds.

4.2   Scope

The fermentation process is employed by the pharmaceutical industry to produce commercially important products that fall under any one of the five categories as shown in Table 4.1.

Table 4.1 Representative examples of some fermentation based products.

Microbial cells or biomass

Yeast cells: used in baking industry
Single cell protein (SCP): used as human or animal food

Microbial enzymes

Bacterial: amylase, protease, glucose isomerase, lipase
Fungal: amylase, protease, pectinase, amyloglycosidase
Animal enzymes expressed in microbial hosts through rDNA technology: streptokinase

---

Microbial metabolites

Primary metabolites produced during log phase of growth/ exponential phase/ trophophase and are essential for the growth of microbe: glutamic acid (flavour enhancer), lysine, phenylalanine, nucleotides, proteins, vitamins, lipids, carbohydrates, citric acid

Secondary metabolites produced during deceleration and stationary phase/ idophase and are not essential for the growth of microbial cells: antimicrobial compounds, enzymes, inhibitors, growth promoters

---

rDNA products

Hormones: growth hormone, insulin
Therapeutic proteins: clotting factors-VIII & IX, interferons

---

Transformation products

Prostaglandins, steroidal drugs

Apart from the above, the fermentative capability of microbes is used for

1. Generation of bioenergy
2. Bioremediation
3. Biomining

4.3   History

Fermentation is commonly termed as traditional biotechnology. It has been used by man knowingly or rather unknowingly since times immemorial. However, if we were to systematically consider the historical developments in fermentation technology, we can start from the ancient times when alcoholic beverages were first produced. Next, we can go on to consider the developments in the production of organic solvents, single cell protein, antibiotics and during the last two decades that of recombinant DNA products. An outline of the historical developments in the fermentation technology is presented in Table 4.2.

4.4   The Fermentation Process

Fermentation has to do with techniques of cultivating microbes which serve as source of industrially and therapeutically useful bioproducts or as a source of mediators of various bioconversions. Now how does one plan and arrange to cultivate the microbes and derive/ recover useful products is simply what is called as fermentation process. Before one starts with a fermentation process, one must first decide upon the type of fermentation that is to be carried out. Different types of fermentation processes are described in the preceding section.

Table 4.2 Historical developments in fermentation technology.

Scientist	Year	Contribution/ development
-	5000BC	production of alcoholic beverages, leavening of
	1700s	bread, retting of flax, production of vinegar
Antony von Leeuwenhoek	1677	built a microscope and observed microbes
Theodore Schwann	1837	proposed that alcoholic fermentation is brought about by living, fungus-like microbes; first observed and described the growth of yeast; stated the need for inclusion of a nitrogen source in culture medium.
Charles Cagniard-Latour	1838	demonstrated that yeast can grow anaerobically; noted that yeast cells do not die on dehydration or freezing
Robert Koch	1881	developed plating procedures for isolation of pure cultures
R. J. Petri	1887	developed a culture dish what we now call 'petri dish
Edward Buchner	1897	demonstrated that enzymes in the yeast bring about alcohol fermentation
E. Wildiers	1901	demonstrated the requirement of a mixture of vitamins or growth factors for the growth and fermentative capacity of yeast

Table 4.2 *Contd….*

Scientist	Year	Contribution/ development
A. J. Kluyver	1924	showed the similarity of metabolic pathways amongst the microbes and higher life forms
Alexander Fleming	1928	discovered (by chance: serendipity!) penicillin
-	1900-40	development of the production of yeast biomass, glycerol, citric acid, lactic acid, acetone, butanol
DeBecze & Liebmann	1944	used air spargers for yeast biomass production
Weizmann	-	established the need for aseptic conditions for fermentation
-	1940 onwards	improvement in aeration, agitation and product recovery methods; improvement of microbial strains; computer aided control of process parameters
-	1960s	development of pressure jet and pressure cycle fermentors development of continuous culture technique
-	1980s	production of recombinant human proteins expressed by genetically modified (GM) microbes

4.4.1 Types of fermentation processes

The fermentation processes may be classified in the following three ways:

1. Depending upon the scheme of fermentation

    (i) *Batch fermentation* : In this case, the sterile nutrient broth is introduced into the pre-sterilized fermentor and inoculated. The culture goes through various cell growth phases like lag, log, stationary and death phase. The desired product and other metabolites accumulate in the fermentor and at the same time nutrients get exhausted. The process is halted/ stopped, desired product is recovered and the fermentor is cleaned and sterilized for next inoculation.

    (ii) *Fed batch fermentation* : This is also a batch process i.e. the fermentation process has to be halted/ stopped at one particular stage, but a number of batches of nutrients are fed during the course of process, instead of just one filling at the start as in batch process. The nutrients are fed just before the culture is likely to go into stationary/ death phase. Then at one stage the process ultimately has got to be stopped to recover the desired product. It is to be noted that unlike the continuous process (discussed next), previously fed culture medium from the fermentor is not removed when new culture is fed. Thus, volume of fed-batch culture increases with time.

(iii) *Continuous culture :* Here, the fermentor is continually replenished with a supply of nutrients and at the same time the desired and metabolic products are recovered from the broth after its simultaneous and continuous removal from the fermentor. This ensures maintenance of the same culture conditions continuously throughout the process over a long period of time.

(iv) *Cascade fermentation :* Here the broth containing the formed product is separated from the biomass (fermenting microbe) and passed on, through a series of fermentors i.e. to say the broth 'cascades' from one to the other fermentor. Each time the collected broth contains more of the formed product. The idea behind 'cascading' is that the microbes (more strictly and correctly the *strains*) used in the subsequent fermentors are such that they are better adapted to work (ferment) in the presence of the high concentration of the product. For example fermentation of beer, wherein broth containing beer is collected and passed on to the next fermentor where there would be a strain of yeast suited to work in presence of high concentration of alcohol. This can be repeated ('cascaded') a number of times to increase the alcohol content in the broth.

2. Depending upon when the product is made

(i) *Type-I fermentation :* Here the desired products are produced during the tropohophase equivalent to the log phase of growth and the products are essential for the growth of microbial cells i.e. these are products of primary metabolism.

(ii) *Type-II fermentation :* Here the desired products are actually made from secondary metabolism while the culture is still growing i.e. undergoing primary metabolism.

(iii) *Type-III fermentation :* Here the desired products are produced during the idiophase, equivalent to the stationary or death phase of growth and the products have no function in cell metabolism i.e. these are products of secondary metabolism.

3. Depending upon how the culture is kept pure

(i) *Aseptic/ sterile fermentation :* This is the most common type of fermentation practiced. This involves sterilizing the fermentor, medium, ancillary equipments and inoculating the pure culture under aseptic conditions i.e. excluding/ preventing the entry of all other microbes.

(ii) *Protected fermentation :* This involves carrying out the process under conditions (not aseptic) which allow and favour the growth of desired (fermenting) microbe while the growth of unwanted (contaminating) microbes is prohibited. For example fermenting strains growing at high temperature, extremes of pH, utilizing hard to metabolize substrates, etc.

(iii) *Consortium fermentation :* This involves growing a group of microbes together, the growth of which is dependent on one another.

4.4.2 Component parts

The component parts (unit operations) of a fermentation process are generally divided into two broad groups:

1. upstream process: which includes all the operations before starting fermentation and
2. down stream process: which includes all the operations after the fermentation has occurred.

Thus a typical fermentation process consists of the following six component parts (or operations):

1. *Upstream process*
   - (i)   Inoculum development
   - (ii) Preparation of culture medium
   - (iii) Sterilization of culture medium

2. *Fermentation*
   - (iv) Control of process parameters

3. *Down stream process*
   - (v)   Product recovery and its purification
   - (vi) Effluent treatment and waste disposal

These six component parts shall now be considered stepwise under separate headings.

4.4.2.1   Inoculum development: Industrially useful microbe

1. Characteristics:

The microbe (or more specifically the strain) that is to be employed for fermentation should desirably possess the following industrially useful characteristics :

- (i)   Carry out the fermentation using a cheap medium/ substrate
- (ii)   Have an optimum temperature for growth and operation above 40°C so that cooling costs are reduced.
- (iii) A consideration of the reaction of the organism with the ferment
- (iv)   Genetic stability
- (v)   Consistency in production
- (vi)   High-yielding for the desired product
- (vii) Permit easy recovery of the formed product
- (viii) Other desirable properties like resistance to infections, non-foaming, morphologically favorable, tolerant to low oxygen tension etc.

2. Source of the industrially useful microbe:

Now, where and how can one get such a microbial strain that possess the above mentioned characteristics? The obvious answer is nature! The soil, oceans, mountains, marine life, deciduous life, the rotting fruits in market, garbage, animal fecal matter, space, or the entire universe for that matter will have to be screened to isolate such a microbial strain! Ones such a microbial strain has been isolated, it can then further be improved. Such a strain isolated from the nature, called the 'wild type' or an improved strain, (which is patentable; life and patentable?!), needs to be preserved so that it can be used industrially in future.

Microbial strains for fermentation may be obtained in one of the following ways:

(i) *Culture collection :* Countries round the world have and supply culture collections (e.g. Microbial Type Culture Collection at Institute of Microbial Technology, Chandigarh) provdes microbes known characteristics. A partial listing of world culture collections is presented in Table 4.3.

Table 4.3 Culture collections of the world: a partial listing.

Culture collection	Abbreviation	Country
American Type Culture Collection	ATCC	U.S.A.
Collection Nationale de Cultures de Microorganisms	CNCM	France
Japan Collection of Microorganisms	JCM	Japan
Microbial Type Culture Collection	MTCC	India
National Collection of Type Cultures	NCTC	U.K.
National Collection of Yeast Cultures	NCYC	U.K.
National Collections of Industrial and Marine Bacteria	NCIB NCMB	U.K

(ii) *Isolation* : As mentioned above, a specimen (most commonly soil) from nature (Universe!) is collected from which the desired microbe needs to be isolated. If considered practically, this task is next to impossible. For this, a technique is required that would allow obtaining and testing of a large number of microbes but without having to perform exhaustive and extensive studies on all the individual microbes. Such a technique is called 'screening'. With rapid advances in molecular biology, genetics and immunology, innovative screening methods have been developed which are based on the use of :

- super sensitive strains as test organisms
- inhibition of enzymes
- binding with receptors
- receptor gene assays
- molecular probes
- ELISA (Enzyme – linked immuno sorbent assay)

(iii) *Improvement*: The natural isolates as obtained above would be the ones that produce the desired product and in the highest possible yield. But still the yield from the natural isolates is considered low to be of any economic importance. Thus programmes are taken up to modify the 'blueprint' i.e. the genetic machinery of the microbe that is responsible for the product formation. This is done by one of the following ways:

- Selection of a natural variant i.e. an unusual strain producing higher yields

- Induction of mutations using mutagens like ionizing radiations (γ radiation), exciting radiation (*u.v.* light), alkylating agents, nitrous acid, purine and pyrimidine analogues etc. These happen to have a lethal effect, killing approximately 90%-99% of cell population. The small proportion of survivors will have mutants i.e. to say not all the survivors are mutants and further out of the mutant population only a few would be desired mutants i.e. mutants having the desired property, say production of higher yield. It is this fraction of cells that need to be selected.

- Genetic recombination: This involves artificially (by man) combining of genes present in different organisms (microbes, plants or animals) to generate an organism having new combination of genes (a patentable organism!). This is being done by one of the following ways:

    - Recombinant DNA technology
    - Protoplast fusion
    - Parasexual recombination in fungi

By using rDNA technique, it has been possible to produce increased yields of enzymes, amino acids, produced by microbes. Further, it has become possible to produce foreign proteins i.e. heterogeneous proteins, such as hGH, insulin, factorVIII etc. in microbes.

3. **Preservation :**

Having obtained/ improved the desired microbial strain, it next becomes important to preserve it such that it remains genetically stable, viable and non-contaminated until used. Microbes are preserved using one of the following ways:

(i) *Lyophilization or freeze drying:* Here the microbial cells are suspended in a medium such a milk, serum or sodium glutamate, transferred to an ampoule, freeze dried (i.e. at low freezing temperature, a high vacuum is applied until water sublimes) and sealed.

(ii) *Dried culture:* For this, the culture to be preserved is inoculated in moist, sterile soil (or silica gel, porcelain beads) and incubated for several days and then dried.

(iii) *Cryopreservation:* For this, microbes, suspended in a cryopreservative agent such as 10% v/v glycerol, are stored at a very low temperature (-150°C to -196°C) in liquid nitrogen refrigerator.

Cultures using the above three methods can be stored for up to 30 years without loss of viability.

(iv) *Agar slope:* Cultures can be stored on agar slopes at 5°C or -20°C in a refrigerator or freezer or Biochemical Oxygen Demand incubator. Such cultures need to be sub-cultured at six monthly intervals which can be extended up to one year if sterile mineral oil is used to cover the slopes.

4.   Development of inoculum :

From the above account, it can be appreciated that cultures are commercially available in the preserved form i.e where life (microbes) are in a state of 'suspended animation'. These cultures are called stock cultures. Therefore, to start out, it is required to 'animate' these cultures, which means to allow germination of the cultures to give vegetative form of life. Further, these vegetative/ maintenance cultures are to be increased in volume from few milliliters to several hundreds of liters which shall be used to 'seed' the medium in fermentor. This process starting from stock culture and going up to the level of 'seed culture' is referred to as inoculum development.

The inoculum should desirably have the following characteristics:

1.   short or no lag phase at all. This can be achieved by-
   - sing an inoculum in the exponentially growing phase
   - using a large inoculum size i.e., 3%-10% by volume of production medium
   - using an inoculum that has been grown in the same or similar medium as the production medium

2.   suitable morphological form

3.   pure culture or that is to say free from contamination and

4.   genetically stable culture that has retained its capability of producing maximum yield of the desired product.

Medium that allows germination of spores is called germination or vegetative or maintenance medium, depending upon the purpose. Alternatively, it is termed as the master culture. From the master culture stage followed by vegetative form, the culture is increased in size which may take several transfers. These subsequent cultures are called seed cultures. The final seed culture serves as the inoculum to be introduced into the sterile medium in the fermentor. The medium in the fermentor is called as the production medium i.e. the medium utilizing which and/or into which the desired products are made and/or secreted.

We shall consider the production medium in section 4.4.2.2. What we are considered about now is the medium for inoculum development i.e. basically seed culture medium. During inoculum development, our aim is to increase the inoculum size i.e. biomass or number of microbes. At this moment we are not interested in product formation, which is desirable during fermentation process using production medium.

With this background in mind, we can now make the following few pertinent points concerning inoculum development:

1. The composition of vegetative/seed culture medium may be different from that of production medium.

2. However, in order to reduce lag time (and therefore the fermentation time) the inoculum development medium should be, as far as possible, similar in composition to the production medium.

3. Similarity of medium composition will also conserve viability and thus productivity.

4. Since a number of stages (transferences) are involved to increase the number of microbes to an optimum level (3%-10% of production medium volume), there happens to be a great risk of firstly, contamination and secondly strain degeneration. This means that over a number of generations the selected natural variant or the artificially created mutant may revert back to the 'wild' or 'normal' type. Therefore, it becomes necessary to perform quality checks on the seed culture to ensure whether the microbial strain still possess the improved quality.

4.4.2.2 Preparation of culture medium

Culture medium is a preparation containing nutrients/ food/ substrates that serve as source of energy and induce the growth of microbes under laboratory/ pilot/ industrial conditions.

The ideal characteristics of a culture medium, suitable to the industrial needs are as follows:

1. It should produce maximum yield and concentration of desired product or the biomass (which ever is desired) per gram of the substrate used.

2. It should give maximum product or biomass formation with the minimal amount of inoculum.

3. It should permit rapid growth of desired biomass (i.e. if biomass is the desired product).

4. It should permit the formation of desired product at a rapid rate.

5. The yield, concentration and rate of undesired product formation should be minimal.

6. It should be of consistent quality i.e. the concentration and rate of biomass or the product formation should be reproducible.

7. It should be readily available round the year.

8. It should be stable and non-problematic during preparation and sterilization.

9. It should not create problems during fermentation process with respect to aeration, agitation, extraction, purification and waste treatment.

10. It should be cheap.

11. A consideration of its physical characteristics especially its rheological behaviour is important.

Culture media/ components may either be simple or complex. The complex media/ components are of natural origin which contain numerous undefined components and in unknown concentrations. Though they are cheap (an important advantage) but are inconsistent in quality as a result of which the growth of biomass and formation of product are unpredictable. Further, the recovery of the formed product becomes a cumbersome task. On the contrary, simple media, wherein the components and composition are known and defined, overcome all the aforementioned disadvantages of natural complex media. But, simple, defined media are less preferred by the industrialists due to their higher cost factor. But actually on the whole, product formation and recovery tend to be higher and easier which actually makes the overall process cheaper.

The microbes can be classified depending upon their source of energy and nutrition. Such a classification is presented in Table 4.4.

Table 4.4 Classification of microbes depending upon source of energy and nutrition.

Source of energy	**Nutrition**	
	Organic	*Inorganic*
Light	Photoheterotroph or Photoorganotroph	Photolithotroph or Photoautotroph
Biological oxidation	Chemoheterotroph or Chemoorganotroph	Chemolithotroph or Chemoautotroph

Now, out of the four types of the microbes, the ones employed industrially are the chemoorganotrophs i.e. ones which utilize reduced i.e. organic form of carbon, nitrogen and derive energy for growth by way of biological oxidation.

One of the major constituents of a culture medium is the carbon source. This influences the rate of biomass formation and / or secondary metabolite production. Many a times it is this carbon source that is actually utilized by the microbe and is converted into the desired product (single cell protein or secondary metabolite). Therefore, a carbon source giving the maximum product yields, one that is cheapest and readily available should be worked out, since the cost of the product greatly depends on it.

Therefore a company/ companies producing the same fermentation product may employ different carbon substrates depending upon its/their geographic location of production site and the prevailing cost of the raw material there. For example, there is an MNC which employs ten different substrates for producing an antibiotic at its production sites round the world. Further, the choice of the carbon source to be selected is influenced by its freedom from metallic impurities, suitability and stability on autoclaving and local government laws. In Table 4.5, presented are the commonly employed media ingredients.

Table 4.5 Various classes of ingredients used in fermentation media.

Ingredient	Comment
1. *Carbon source*	
Carbohydrates	
Starch	Obtained from maize, cereals and potatoes
Ground maize and cereals	-
Barley malt	Obtained from heat treated, partially germinated (by soaking in water, so that starch is broken down and thus becomes easier to assimilate) barley grain
Sucrose	Obtained from sugarcane/beet juice
Molasses	This is the residue (mother liquor) still having 54 % w/v of sucrose left after crystallization of sucrose from sugar cane/ beet juice. This is the most commonly used and cheapest substrate but contains many impurities thereby making recovery of the product complicated and expensive.
Lactose	Of limited use; a side product of dairy industry
Whey powder	It is the water separated from milk/ curd and converted to powder by drying
Cellulose	A potential raw material but of limited use since only a few microbes can degrade it
Corn steep liquor meal }	Used as primary source of nitrogen but also provide appreciable Soyabean amounts of carbon
Sulphite liquor	A byproduct of paper pulp production; contains much of fermentable sugars from wood and not much of cellulose
Oils anf fats	
Vegetable oils	Oils obtained from olive, maize, cotton seed, linseed, soybean etc. containing fatty acids. In comparison to the carbohydrates mayactually be cheaper and provide higher amount of energy.
Glycerol trioleate	-
Methyl oleate	The sole carbon source in cephalosporin production
Hydrocarbons	
n-alkanes Methane Methanol Natural gas	-
2. *Nitrogen source*	
Inorganic	
Ammonia gas Ammonium salts }	Used also to produce an acidic pH
Nitrates	Used also to produce an alkaline pH

Organic	
Amino acids	Few mutant microbes have an absolute requirement for certain amino acids
Proteins	-
Urea	-
Complex organic nitrogen sources	e.g. soybean hydrolysate, corn steep liquor*, soy meal, peanut meal, cotton seed meal, distiller's solubles, meat extract, yeast extract, pharmamedia, peptones, casein hydrolysate; they also provide many amino acids, vitamins and minerals. * A liquid containing low molecular weight sugars and peptides, produced by submerging maize grains in water to soften before wet milling for extraction of starch
Chemically defined amino acid media	For vaccine production
3. Minerals	
Macroelements	
Magnesium	-
Potassium	
Sulphur	
Phosphorous	As inorganic phosphate used as buffer; affects target enzyme involved in production of antibiotics
Calcium	Precipitates out excess of phosphate
Chloride	Required by halophilic bactria; enhances the yield of chlorine containing metabolites such as chlortetracycline, griseofulvin, caldriomycin etc.
Trace elements	
Validium,	Most commonly the concentrations of manganese, iron and zinc affect primary and secondary metabolism;
Chromium,	
Manganese, iron,	
cobalt, nickel,	Atomic nos 23-30
copper, zinc,	
molybdemun	Atomic no. 42
4. Growth fators*	
* These are preformed compounds (i.e. cannot be synthesized by microbes) and must necessarily be added to the medium:	
Vitamins	
B-complex group	-
Calcium pantothenate	For vinegar production
Biotin	For glutamic acid production
Amino acids	Provided along with complex organic carbon or nitrogen sources
Fatty acids or sterols	
5. Chelators	
EDTA	Eliminates the problem of precipitation of metal ions with citric acid phosphates
Polyphosphate	

Table 4.5 Contd....

6. *Buffers*

Calcium carbonate phosphates	Maintain the pH at about 7.0 (the most common pH for many industrial media)

7. *Metabolic regulators**

*These help to regulate the production so that more of a desired product is produced:

Precursors	*For preferentially increased formation of:*
Phenyl acetic acid	Penicillin G (i.e. Benzyl penicillin)
Phenyl ethylamine	
Phenoxy acetic acid	Penicillin V
Chloride	Chlortetracycline
Cyanide	Vitamin B_{12}
β-ionone	Carotenoids
α-amino butyric acid	L-isoleucine, Cyclosporine-A
Anthranillic acid	L-Tryptophan

Inducers	*Especially used for inducing the growth of following enzymes:*
Starch, dextrin	Amylase
Maltose	Pullulanase
Pectin	Pectinase
Isovaleronitrile	A very potent inducer of nitralase
Fatty acids	Lipase
Yeast mannan	In cultures of *Sreptomyces griseus* for production of streptomycin, induces production of β-mannosidase which converts mannosidostreptomycin (an undesirable product formed along with streptomycin) to streptomycin
Galactomannan	β-galactomannosidase
Cellulose	Cellulose
Proteins	Proteases

Inhibitors	*Inhibit/suppress the formation undesired product:*	*So that more of the desired product is formed:*
Sodium bisulphite	Acetaldehyde	Glycerol
Bromide	Chlortetracycline	Tetracycline
Alkali metal, Phosphate and pH below 2.0	Oxalic acid	Citric acid
Di-ethyl barbiturate	Various other rifamycins	Rifamycin B

8. *Antifoaming agents**

* These are surfactants i.e. they reduce the surface tension in foams, break them and also destabilize protein films:

Alcohols	E.g. Stearyl alcohol
Esters	
Fatty acids	E.g. Glycerides, vegetable oils, cod liver oil
Silicones	
Sulphonates	
Polypropyleneglycol	

9. *Water*

Deionized water	This is the major component of any medium; consideration should be given to its pH, dissolved salts and effluent contamination

4.4.2.3   Sterilization of culture medium

Why do we talk of and lay so much of emphasis on sterilization in a fermentation process? Just consider the consequences if unwanted (contaminating) microbes are present in the fermentor vessel. It can result into any one of the following:

1. Substrates in the medium will be consumed by the contaminating microbes resulting in loss of productivity
2. The contaminating microbes may outgrow the desired microbe especially so in a continuous fermentation process
3. The desired product (metabolite or biomass for single cell protein) would be contaminated
4. The extraction of the desired product would become difficult and expensive
5. possibly the desired product can get degraded by the action of the contaminating microbes e.g. contamination by β-lactamase producing bacteria in penicillin fermentation
6. Contamination by bacteriophages! Oh no, it will lyse the entire bacterial culture.

Therefore, it becomes necessary to sterilize the medium and fermentation vessel, to use pure inoculating culture and to follow aseptic technique. At this point, refer section 4.4.1. You should keep in mind that fermentation process may be of 'protected type' i.e. not requiring strict conditions of sterility and asepsis (e.g. production of beer).

We shall first consider the methods for sterilizing the culture medium and then methods adopted for sterilizing the fermentor.

1. **Sterilization of culture medium:**

    Microbial and plant culture media are practically sterilized by moist heat using steam, while animal culture media, since they contain thermolabile components, are sterilized mechanically by way of filtration.

    However, while using moist heat, it is to be noted that substrates having carbonyl groups such as reducing sugars react (Millard-reaction) with compounds having amino groups such as amino acids, urea etc. resulting in discoloration (browning) of the medium. Therefore, the two substrate solutions should be sterilized separately and aseptically combined after cooling. Further, it is to be noted that vitamins, amino acids and proteins are liable to degradation by steam sterilization. Practically this problem is overcome by exposure of medium to a high temperature for short duration of time. Such a regimen proves more lethal to the contaminating microbes and is at the same time less damaging for the nutrients.

    Sterilization of medium is achieved industrially by one of the following means:

    (i) *Batch process :* in this case the medium is filled in the fermentor and sterilized *in-situ* or alternatively the medium is sterilized in a specially designed, separate vessel called 'mash cooker' and then fed into the fermentor.

2. ***Continuous process :*** In this case the medium flows continuously through heat exchangers. For ease of understanding, the process can be understood to occur in three phases as follows:

- *Heating up period :* Medium achieving sterilization temperature,

- *Holding period at sterilization temperature :* Medium actually flows at predetermined rate (volume of medium in a given period of time) through insulated serpentine coils held at the sterilizing temperature and

- *Cooling period :* Medium is cooled down to the fermentation temperature.

With this continuous method, a high temperature and short holding time can be achieved.

Likewise, the feed i.e. additives administered during the fermentation process also can be sterilized.

2. **Sterilization of fermentor vessel :**

This is required if the medium has been sterilized separately by batch process or has to be sterilized by continuous process.

For sterilization of fermentor, live steam is introduced in the vessel, held at 15psi for 20 minutes. At the end of the process, sterile air is purged in and the vessel held at positive pressure to prevent the entry of extraneous, unsterile air.

All that we have just considered, concerning the industrially useful microbial strain, is summed up in a nut shell in figure 4.1.

4.4.2.4   Fermentation

A fermentor consists of the following four components:

(i)	the fermentor hardware or vessel	→ physical component
(ii)	the sensors that maintain the system	→ physical component
(iii)	the culture medium	→ chemical component
(iv)	the microbial strain	→ biological component

Recently one more component has been added i.e. the computer control system. The medium and organism have already been considered. In this section we shall focus our attention on to the design of fermentor vessel and the process parameters that are monitored.

SOURCE

- Procurement of desired strain from culture collection from anywhere round the world

- Isolation of natural variant from soil etc. using screening technique

- Improvement of a wild, type using:
1. Mutation and selection program
2. Genetic recombination by-
   i. Parasexual recombination
   ii. rDNA technique
   iii. Protoplast fusion

PRESERVATION

- Cryopreservation using liquid nitrogen at -150°C to -196°C

- Freeze drying- sublimation

- Drying a soil culture

- Storing on agar slopes at 5°C or at -20°C covered with mineral oil

DEVELOPEMENT

Stock/Master culture

(Spores lyophilized with skim milk/lactose: microbes in 'suspended animation')

↓

Vegetative/Germination culture

(Microbes in vegetative form: 'animated')

↓

Maintenance culture

(Agar slope incubated every half to weekly intervals at 28°C)

↓

Seed culture

(One to few or several stages/transferences: to increase the inoculum volume from few ml to 10's or 100's of liters)

↓

Inoculum

(Final seed culture: 3%-10% by volume of production culture)

↓

Production culture

(100's to 100 thousands of liters)

Figure. 4.1 Developement of the industrially useful microbial strain.

1.   **Fermentor : its design**

The points pertinent to the design and construction of a typical fermentor include the following:

(i)   The vessel should be such as to prevent the entry of extraneous microbes into the medium i.e. it should be operable aseptically

(ii)   At same time, vessel should be such as to prevent the escape of microbe employed for fermentation out of the vessel (known as containment), since the microbe may be hazardous to the health of workers or society

(iii) There should be adequate provision for-

- aeration,
- agitation,
- temperature control
- pH control,
- sampling facilities,

(iv)   Should have low requirement for power

(v)   Should have minimal evaporation losses

(vi)   Its operation, harvesting, cleaning and maintenance should be easy and labour non-intensive

(vii) Should be suitable for a range of processes

(viii) Should have smooth internal surfaces

(ix)   Should have a geometry similar to the vessels used for pilot plant or scale up stage

(x)   Its material of construction should be cheap

(xi)   There should be adequate service provisions such as compressed/sterile air, hot/cold water, steam/ condensate, electricity/ generator, drainage/ storage facilities, monitoring/ maintenance equipment, extraction/recovery equipment etc.

For small scale/laboratory use, fermentors made of glass are commercially available in capacities ranging from 1-50 liters while for pilot-scale and for industrial use, fermentors are made of stainless steel and their capacities range from 50 liters -1000 liters and 1000 liters - 1,000,000 liters respectively.

A design of a typical fermentor and a consideration of its dimensions are shown in Figure 4.2 and Table 4.6 respectively.

2.   **Agitation and aeration in fermentors :**

The contents of a submerged fermentor (suspended growth system) have to be agitated for the following reasons:

(i)   Mixing of bulk liquid and gas phases

(ii)   Rapid dispersion of air

(iii) Rapid mixing of introduced material

(iv)   Transfer of oxygen

(v)   Suspension of particulate solids and

(vi)   Maintenance of a uniform physical, chemical and biological environment inside the vessel

Table 4.6 Geometrical ratios of a fermentor having three multi-bladed impellers.

Dimensions	Typical value/ ratio
Operating volume	170liters
Liquid height	150cm
Tank diameter (L/D)	1.7
Impeller diameter (P/D)	0.33
Baffle width/D	0.098
Impeller height (H-Z)/D	0.37
P/V	0.74
P/W	0.77
P/Y	0.77
P/Z	0.91
H/D	2.95

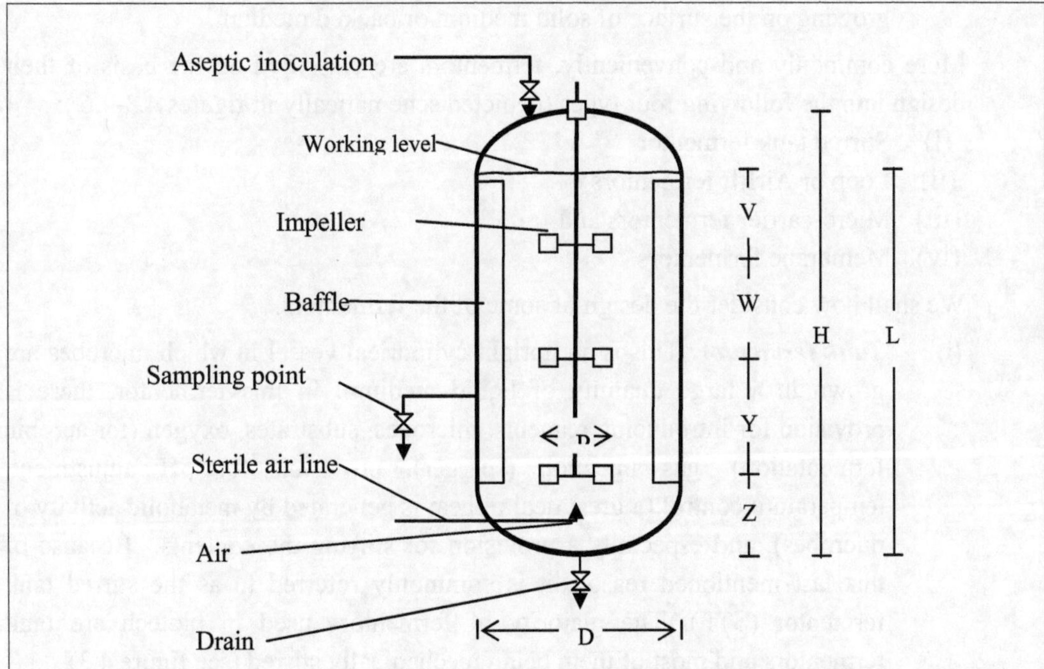

Figure. 4.2 Design of a typical fermentor: An upright cylindrical vessel.

The agitation in fermentors is achieved by one of the following ways:

(i) *Mechanical agitators:* Impellers e.g. disc turbine, vaned disc, open turbine, marine propeller

(ii) *Forced convection:* In this, liquid medium from the fermentor is circulated by pumping through a gas entertainer and then back into the fermentor.

    (iii) *Sparger/Gas lift or Air lift/Aeration system:* A sparger or a diffuser is a pipe or plate with holes in it which is used to introduce air into the liquid medium in the fermentor. It is located at the base of the fermentor and when compressed gas is admitted through this porous device, bubbles shoot up over a wide area at the base of the tank. This bubbling provides for both agitation and aeration as well.

3.   Types of fermentors:

Fermentors may be classified into various types depending upon different schemes. But generally, they are grouped into two types:

    (i)   Submerged fermentors or suspended growth systems i.e. where the cells are dispersed in the liquid medium

    (ii)   Surface fermentors or supported growth systems i.e. where the cells are growing on the surface of solid medium or packed medium.

More commonly and conveniently, fermentors are classified on the basis of their design into the following four types (depicted schematically in figures 4.3-4.6.):

    (i)   Stirred tank fermentor

    (ii)   Loop or Airlift fermentors

    (iii)   Micro-carrier fermentors and

    (iv)   Membrane fermentors

We shall now consider the design of some of the fermentors.

    (i)   *Tank fermentor:* This is an upright cylindrical vessel in which microbes are grown in a large quantity of liquid medium. In this fermentor, there is provision for introducing reagents, microbes, substrates, oxygen (for aerobic fermentation), gas injection (sparger), provision for pH adjustment, temperature control (a great deal of heat is generated by metabolic activity of microbes), and especially a provision for stirring the contents. Because of this last mentioned reason, it is commonly referred to as the stirred tank fermentor (STF). The majority of fermentors used in biotech are tank fermentors and most of them being mechanically stirred (see figure 4.3).

    (ii)   *Air lift fermentor:* These belong to the class of loop fermentors. The vessel consists of two parts: an up flow tube or riser and down flow tube of downcomer (see figure 4.4). It is suitable for use with liquid fermentation medium. The medium in the riser is driven up by the bubbles of a gas (air or oxygen) generated by pumping it into the bottom through the sparger. At the top of the vessel there is a gas separator that disengages the gas from the liquid. The separated gas exits from the vessel while the liquid moves down the downcomer.

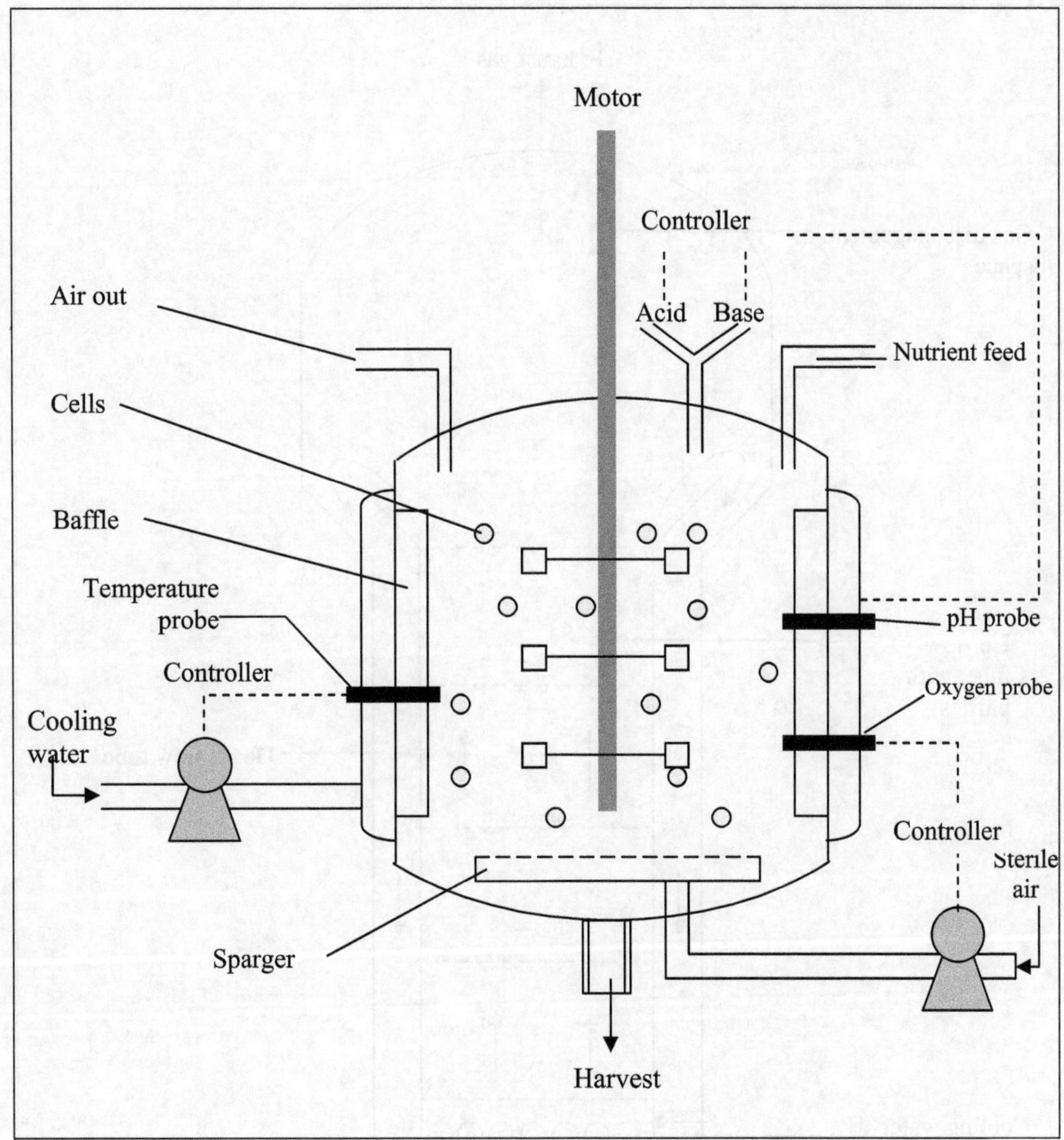

Figure. 4.3 A typical representation of stirred–tank fermentor.

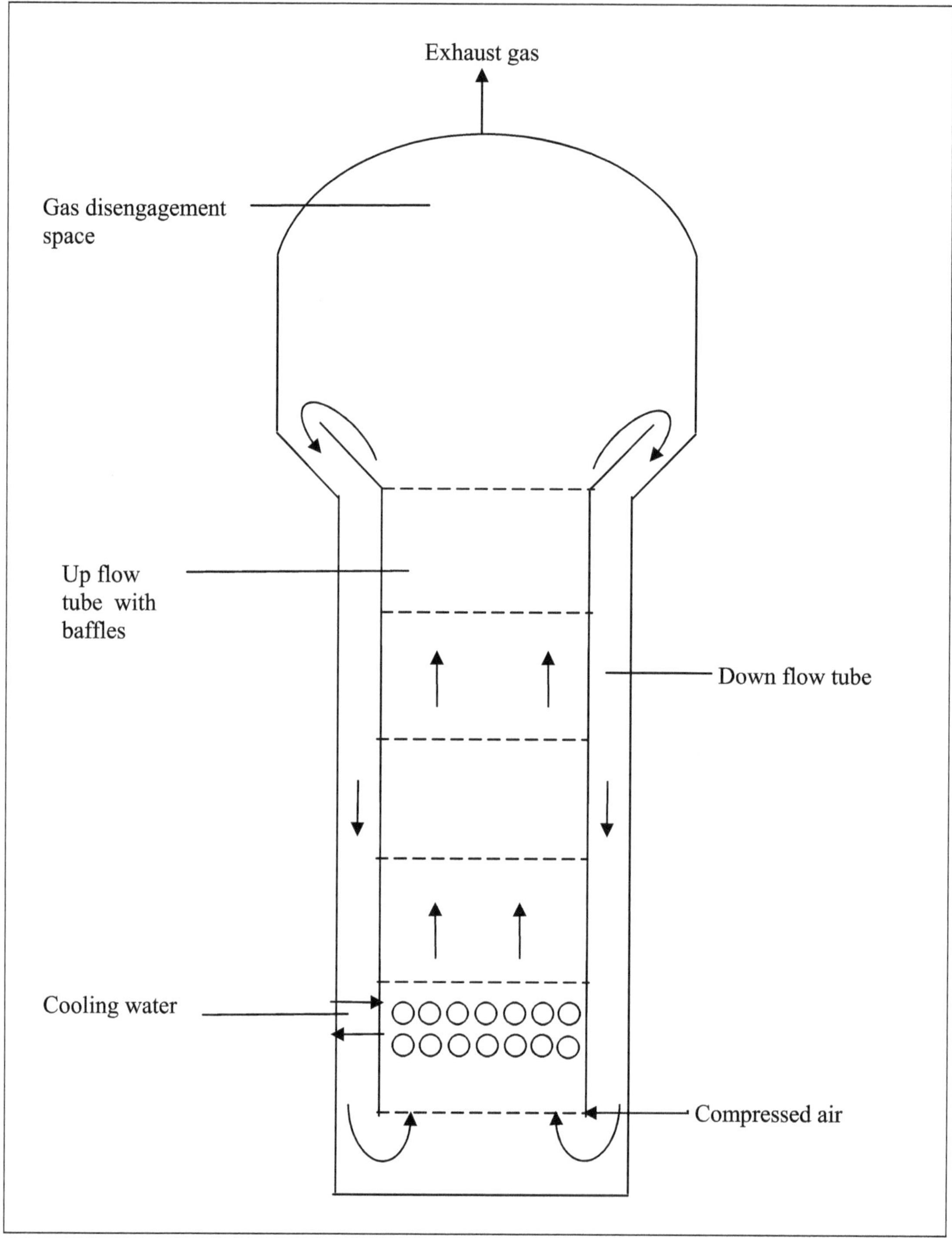

Figure. 4.4 A typical representation of air lift fermentor.

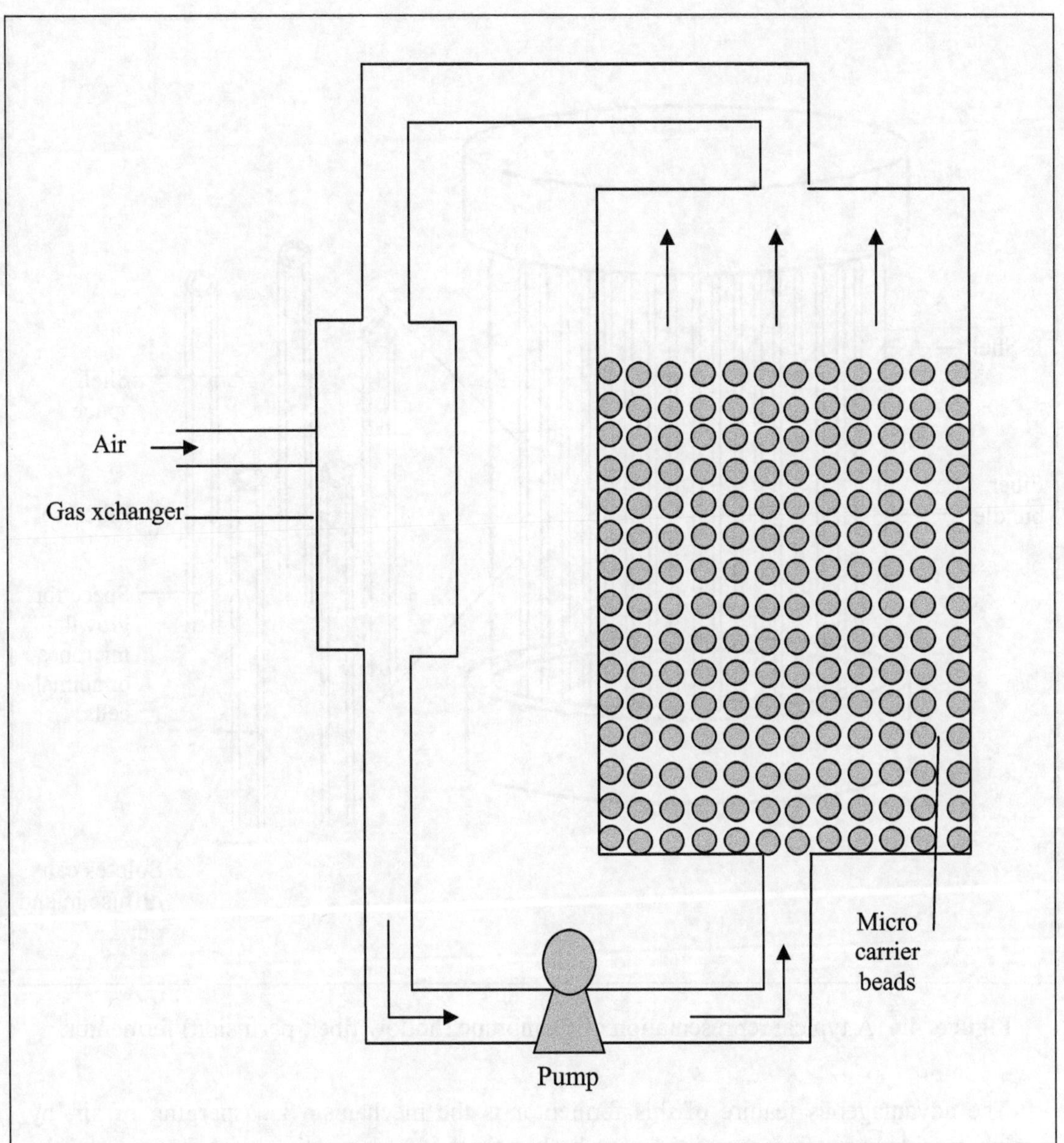

Figure. 4.5 A typical representation of micro-carrier (fixed-bed) fermentor.

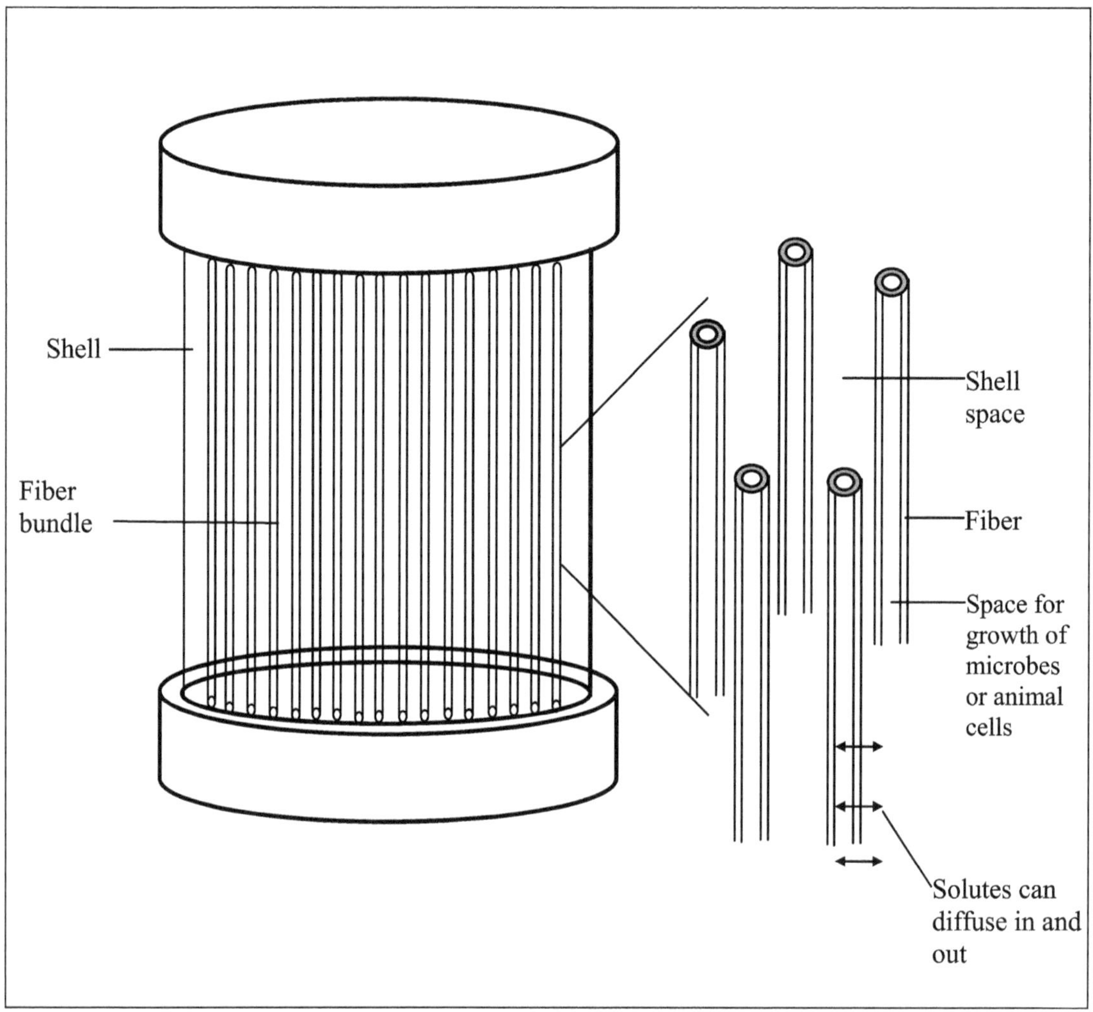

Figure. 4.6 A typical representation of membrane (hollow fiber, perfusion) fermentor.

The advantageous feature of this fermentor is the mechanism i.e. sparging of air, by which liquid circulation, providing for both agitation and aeration, is achieved. In the mechanically stirred fermentors (based on impellers), the paddles, blades or plates of the propellers or turbines, harm the delicate membranes of the mammalian cells or damage the hyphae of fungi.

(i) ***Bubble column fermentor***: In this fermentor, the cells are kept suspended in liquid medium by means of bubbles rising through a tube. Unlike airlift fermentors, the bulk of the liquid does not circulate, the bubbles merely keep the culture cells aerated and suspended (just like a fluidized bed; Wurster process).

(ii) *Immobilized cell fermentors:* These fermentors are preferred for culture of animal and plant cells which get damaged by the stirring action of impellers in STF. Also when the cells are immobilized, it becomes easier and economical to separate them from the substrates and products. Further, animal cells are anchorage dependent i.e. they tend to grow and produce well when they are attached/grow on a support. Examples of immobilized fermentors include: membrane fermentor, filter or mesh fermentor and carrier particle fermentors.

(iii) *Carrier particle fermentor:* In this type, the cells are 'immobilized' onto a solid support such as nylon or gelatin beads. The size of the beads is just about the size of the cells. If the particles are of neutral density, they can be pumped with the liquid and if they have a high density then they can be fluidized or simply they can form a solid bed (see figure 4.5).

(iv) *Membrane or hollow fibre fermentors:* Hollow fibres are tubes made of a porous material. The fibre tubes are very small having an internal diameter of fraction of millimeters, which means these fermentors have a very large ratio of surface area to volume. These fermentors work on the principle of dialysis i.e. a membrane allowing small molecules to pass through it (such as substrate molecules, products, metabolites) while restricting the passage of large molecules (such as microbes). Thus cells are kept inside the fibre tubes while the culture medium circulates (perfuses) outside the tubes.

The hollow fibre fermentors are most commonly employed for culturing animal cells for the production of monoclonal antibodies. The animal cells grow on the inside surface of the fibres, the nutrients diffuse in from outside i.e. from the shell space (see figure 4.6) and the product diffuses out into the shell space.

It is to be noted that this fermentor is unsuitable if the desired product happens to be biomass.

(i) *Digestors :* These are fermentors used for the break down and disposal of sewage or agricultural waste using metabolic capacity of microbes. The digestors are of two types: aerobic digestors and anaerobic digestors. The anaerobic digestors are more commonly used wherein the waste material is packed in an air-tight vessel and the microbes degrade the material accompanied with the generation of methane.

(ii) *Pond fermentors :* Their name is self explanatory. These are simply large ponds used mainly for growing algae (biomass). Because they allow rapid transfer of oxygen (hence also known as oxidation ponds), they are used for reducing the BOD (biochemical oxygen demand) of organic waste.

(iii) *Tower fermentors*: These are elongated, non-mechanically stirred fermentors with a height: diameter ratio (called as aspect ratio) of at least 10:1. In these, the medium is introduced at the base and the formed product is collected from the top. They are suitable for anaerobic fermentation process such as brewing.

(iv) ***Chemostat:*** It is a closed culture vessel for continuous operation. The characteristic feature of chemostat is that the new culture medium is continuously fed in and old medium and microbes are continuously removed from it at fixed rates. Thus, concentration (chemo) is held constant (stat).

(v) *Auxostat*: In an auxostat, unlike cheomostat, the rates of addition of new culture medium removal of old medium vary. The addition and removal of medium are dependent on some feature (auxo) of the culture such as pH or turbidity which is monitored to be held constant (stat). Thus there are pH auxostat (pH held constant by addition of alkali or acid) and turbidostat (turbidity held constant by addition of new culture) fermentors.

With the inclusion of computer component in the fermentation technology, it has become possible to continuously measure and control the concentrations of materials in the fermentor through out the entire process. Thus industrially, it is the turbidostat that is preferred and used over the chemostat.

(d) Process parameters:

In fermentation, there are a number of process variables that need to be monitored and controlled/adjusted through out the course of fermentation. In order to analyze the physical, chemical and biological environments inside a fermentor, large fermentors have 1000's of sensors/probes which are in turn linked to 100's of control valves. What is important and interesting to note is that the environments inside the fermentor need to be analyzed 'on-line' i.e. there and then itself. It could be suicidal to collect out a sample and analyze it in the laboratory over a period of time or i.e. to say analyze it 'off-line'. This is because small alterations in physical or chemical environments would arrest the activity of or can even be fatal to the biological environment i.e. the microbes.

The various process parameters that need to be monitored and adjusted are tabulated in Table 4.7.

Table 4.7 Fermentation process parameters.

Physical	
Temperature	Power input
Pressure	Rheology
Current flow	Amount of mixing
Agitation speed	Foamimg
Chemical	
pH Nutrient concentration	
Redox potential	Product concentration
Gas analysis for CO_2, O_2	Metabolite concentration
Biological	
State of culture	Cell concentration/density

It was mentioned earlier that the computer component has been included in fermentors. The sensors generate a signal which after conversion to a digital form, is fed into a computer for analysis of the parameter. The information concerning a parameter is then used to appropriately adjust the parameter to the preset level. Even the necessary action required to achieve the preset level for the concerned parameter, can be brought about by the computer itself. For this, the computer sends a digital signal back to the controlling valves which on receipt of the signal act accordingly to achieve the preset value.

4.4.2.5   Product recovery and its purification

Downstream process refers to all the operations involved after fermentation has occurred resulting in the formation of the desired product. The operations include recovery of the desired product and treatment of effluent/ waste before it can be discharged. Former operation is considered in this section.

Mostly the desired biotech based products are produced in meager amounts ranging from as low as 10mg-200mg/l to as high as 500mg-800mg/l. Therefore, recovery and purification of the desired product is a difficult, gigantic (involves handling of a large volume of broth), critical and costly step. In fact the higher cost of some of the biotech based products is due to their high recovery costs which may be as high as 70% of total manufacturing cost!

The recovery of a product is carried out in three steps:

1. *Separation*: This involves separation of the crude product from other constituents of broth such as biomass, substrates, metabolic waste products etc. This is most commonly achieved by means of filtration and/ or centrifugation to remove microbial cells, debris, fragments, particles (virus).

2. *Concentration*: This involves removal of most of the broth water and is commonly referred to as dewatering. This is important so that the handling volume is reduced for the next step. During this step, most of the bulk contaminants are also removed. It is brought about by one of the ways presented in Table 4.8.

Table 4.8 Methods for concentration of broth.

Method	Remark
Sedimentation	for large cells
Centrifugation	for small cells
Flocculation	by adding flocculating agents, cells can be clumped together which later settle out
Flotation	this means cells riding on bubbles and rising up as the bubbles move upwards to form foam
Reverse osmosis	based on molecular size of molecules
Ultrafiltration	based on molecular size of molecules
Ion-exchange methods	based on charge of molecules
Adsorption on solids	based on surface area
Solvent extraction	based on hydrophilicity or lipophilicity of molecules
Distillation/fractionation	based on volatility of molecules

3. ***Purification***: This involves isolating the product in its purest form possible, free from all trace contaminants and variant forms of molecules if any.

   It is to be noted that enzymes are used in a concentrated, crude form, if and only if required in pure form then only further purification of enzymes is carried out.

   A term called 'harvesting' is used, when the desired product is the biomass/ cells and the cells are purified.

   Purification can be achieved by one of the ways presented in Table 4.9.

4. ***Cell disruption:*** If the desired product remains inside the cell (most proteins, produced by genetic engineering, enzymes and large molecules), then it becomes necessary to first disrupt the cells to recover the product. Disruption of the cells is achieved by one of the methods given in Table 4.10.

Table 4.9 Methods for purifications.

Method	Remark
Precipitation	by salting out e.g. precipitation of proteins by ammonium sulphate by addition of organic solvent so as to alter the dielectric property of solution e.g. precipitation of dextran or proteins by addition of alcohol
Crystallization	-
two-phase	involves addition of an immiscible liquid (benzene/ ether/ oil and liquid-liquid water) which preferentially dissolves out the product over the separation impurities
Two-phase aqueous extraction	both the phases are aqueous
Polymer addition	e.g. precipitation of proteins by binding with PEG
Polyelectrolyte addition	e.g. precipitation of proteins by triazine dyes
Heat denaturation:	to coagulate the unwanted protein if the desired product is non-proteinaceous or is a 'thermostable' protein
Isoelectric pH	involves addition of acid/alkali to give an isoelectric pH at which the solubility of the desired protein is least and it precipitates out
Dialysis	involves the use of semi permeable membrane for washing away small molecular weight impurities from the desired large molecular weight product
Electrodialysis	involves the application of electric current to hasten dialysis for removal of trace ionic impurities
Supercritical fluids	these are gases compressed to give liquids above the temperature at which they can form liquids; the capacity of these fluids to dissolve chemicals is dependent on their pressure and temperature, by varying which almost any thing can preferentially be dissolved
Chromatography	this is of various types like affinity, gel filtration, ion-exchange, hydrophobic and reverse phase

Table 4.10 Methods for cell disruption.

Mechanical method	
Using	*Remarks*
High pressure homogenizers press (Liquid shear)	involves forcing the suspension of cells (French through a small hole at high pressure
Mills	involves the use of abrasives, metal balls, (ball mill) or rods (cylinder mill)
Blenders	i.e. food processors
Solid shear (Hughes press)	frozen microbes being forced through a small orifice

Autolysis	
-	involves alteration of cell condition so that the cell digests itself

Enzymatic method	
Using	*Remarks*
	involves the use of enzymes that dissolve away the rigid cell wall such as
lysozyme for bacterial cell wall	
chitinase for fungal cell wall	
gluconase for fungal cell wall	
cellulose for plant cell wall	

Chemical method	
Using	*Remarks*
Detergents	treatment of cells with chemicals results in disruption of plasma membrane and other lipoidal layers
Quaternary ammonium compounds	
Sodiun lauryl sulphate	
Sodium dodecyl sulphate	
Triton X-100	
Alkali	
High salt concentration	disruption by such method is plasmolysis
Organic solvents	
Pure water	disruption by such method is osmotic shock

Freeze thaw method	*Remarks*
-	involves freezing the cells which result in solidification of internal structures having water and then thawing so as to break up the internal structures having ice crystals

Ultrasonication	*Remarks*
-	involves the use of ultrasound waves for cell disruption

4.4.2.6   Effluent treatment and waste disposal

Having recovered the desired product (only 5%-10% of the total broth volume!), what ever is left over is waste! This includes unconsumed and/or unconverted inorganic and organic substrates, microbial biomass, cell disintegrates, water (also from cleaning, cooling operations), various kinds of suspended matter etc. This (biotech) fermentation waste contains a considerable amount of organic matter. Now, if this is allowed to run down into the nearby water reservoir (river, lake etc.) then the aquatic- microbial flora utilize the oxygen dissolved in the water to oxidize the organic matter. This results in lowering of the dissolved oxygen concentration in the water region where effluent is run down. The consequence, as is obvious, is the death of certain oxygen dependent species (fish for example) in that particular region. At the same time certain species capable of growing at lower oxygen concentration will thrive. This phenomenon in ecological sciences is termed eutrophication. The amount of oxygen needed by the microbes to oxidize the organic matter is termed as biochemical oxygen demand (BOD). While, chemical oxygen demand (COD) refers to the determination of oxidizable matter chemically in waste by treating it with boiling acidic potassium dichromate solution for two and a half to four hours.

Therefore, the aim in effluent treatment is to reduce the amount of organic matter i.e. to reduce the BOD of waste. Secondly, the phosphates and nitrates are also to be removed. Consider this: Domestic sewage has a BOD of 350mg/l.; while in penicillin production, the wet mycelium from filter has a BOD of 40,000 to 70,000mg/l.

There are quite a number of ways to treat the waste material, some of which are presented in Table 4.11.

Table 4.11 Methods for effluent treatment.

Physical	
Dewatering by filtration	Incineration of solid waste
Sedimentation	Solid waste being composted
Centrifugation to remove solids waste	Solid waste being used as fertilizer
Chemical	
Coagulation	
Microbiological	
Digesting	Biofiltration
Oxidation by microbes	
Remedial	
Pumping in oxygen equal to BOD in the water reservoir to help microbes oxidize the organic matter without causing imbalance!	

4.5   Fermentative Production of Some Products

The fermentative production of antibiotics, vitamins, enzymes and amino acids shall be dealt with in this section. It shall be useful for the reader to refer section 4.4.2, dealing with the component parts of fermentation process, since the fermentative production for the mentioned category of products is being described based on those steps.

4.5.1 Penicillin

Upstream Process

(i) *Inoculum development:*

- The microbe: The microbe employed for the production of penicillin at the industrial level is *Penicillium chrysogenum*. The mould initially discovered by Alexander Fleming was *Pencillium notatum*. Since then the penicillin producing mould has been developed by recombination and/ or mutation by exposure to mutagens especially nitrosoguanidine.

- Vegetative medium: A typical vegetative culture medium for *Pencillium chrysogenum* is given in Table 4.12. It is started as a shake-flask culture which gradually increases in volume. Aeration and agitation of the culture are must.

Table 4.12 Vegetative culture medium for *Penicillium chrysogenum*.

Ingredient	Concentration
Corn steep liquor	3.5%
Sucrose/ glucose	2.0%
Calcium carbonate	0.5%-1%
Inorganic salts	-
Distilled water	qs

2. ***Preparation of culture medium:*** A typical culture medium that is employed for the production of penicillin is given in Table 4.13.

Table 4.13 Production culture medium for penicillin.

Ingredient	Concentration	Use
Corn steep liquor	4%-5%	nitrogen source
Lactose/ Glucose*	3%-4%	carbon source
Phenylacetic acid/	0.5%-0.8%	Precursor for penicillin G
Phenoxy acetic acid	"	Precursor for Penicillin V
Lard oil/ Vegetable oil	0.5%	Anti foaming agent
Calcium carbonate	1.0%	Buffer
Potassium dihydrogen phosphate/ Phytic acid	0.4%	Phosphorous source (250g/ml-500g/ml)
Ammonium sulphate	-	Nitrogen and sulphur source
Micro- & macro-elements	-	Growth and product formation
Distilled water	qs	Solvent

The pH o the production culture medium is to be adjusted to 5.5 -6.0.

3. ***Sterilization:*** Medium is sterilized using steam under pressure.

Fermentation

4. ***Process control parameters/ characteristics:*** The following are notable points concerning the production of penicillin:

   - Production is carried out by submerged culture method
   - Production volume is 40,000 liters-200,000 liters
   - Aerobic process; aeration rate is 0.5-1.0 volumes of air/ minute
   - Agitation is achieved by employing turbine type impellers
   - Power requirement is of the order of 1-4W/l
   - Temperature should be controlled between 23°C-25°C
   - During the process pH rises from 5.5-6.0 to 7-7.7.5. pH of 7.0-7.5 is optimum for production of penicillin.
   - The pH rises due to two reasons: one, the acids supplied through corn steep liquor are utilized and two, the amino acids, also supplied through corn steep liquor, are deaminated to liberate ammonia.
   - At the end of process, pH rises to 8.0.
   - * Glucose in the medium is utilized rapidly to provide growth of mycelia. But penicillin production occurs under glucose starvation conditions. Thus if lactose is used as the carbon source, it hydrolyses slowly to glucose and galactose. The slow glucose availability creates the required starvation conditions and allows for penicillin production.

        Alternatively, if glucose alone is used as the carbon source (since it cheap) then it is added slowly to the medium to create starvation conditions.
   - The sugar (lactose) concentration falls from 3.5% w/v to almost negligible. The depletion of lactose results in autolysis of the mycelium and marks the end of the fermentation.
   - It is seen that during the course of fermentation, the increase in biomass is very small. This is so desired because the nutrients (sugar etc.) are to be utilized to produce penicillin and not the biomass.
   - Time period for completion i.e. when maximum yield is obtained is 5-6 days.

Downstream process

5. ***Recovery and purification:*** There are two methods; one is carbon method which is obsolete and the other that is used industrially is the solvent extraction process. Steps involved in the solvent extraction process include:

   - Filtration using rotary vacuum filters for removal of mycelium from broth
   - Acidification using phosphoric or sulphuric acid to a pH of 2-2.5 for conversion of penicillin to ionic form
   - Extraction using organic solvents such as amyl acetate, methyl isobutyl ketone or butyl acetate in Podbeilniak counter current solvent extractor
   - Treatment with activated charcoal for removal of pigments by adsorption
   - Filtration using rotary vacuum filter for removal of charcoal with adsorbed impurities
   - Basification using potassium or sodium hydroxide for formation of penicillin salt

- Extraction of salt in water
- Acidification of the aqueous solvent resulting in precipitation of penicillin
- Crystallization of pure penicillin as potassium or sodium salt
- Drying using horizontal belt filter for stabilization

Yield that is obtained using *Pencillium chrysogenum* is about 160 times than that produced using the original strain (*Pencillium notatum*). A flow chart for the fermentative production of penicillin is presented in Figure 4.7 and time course of fermentation is depicted graphically in Figure 4.8.

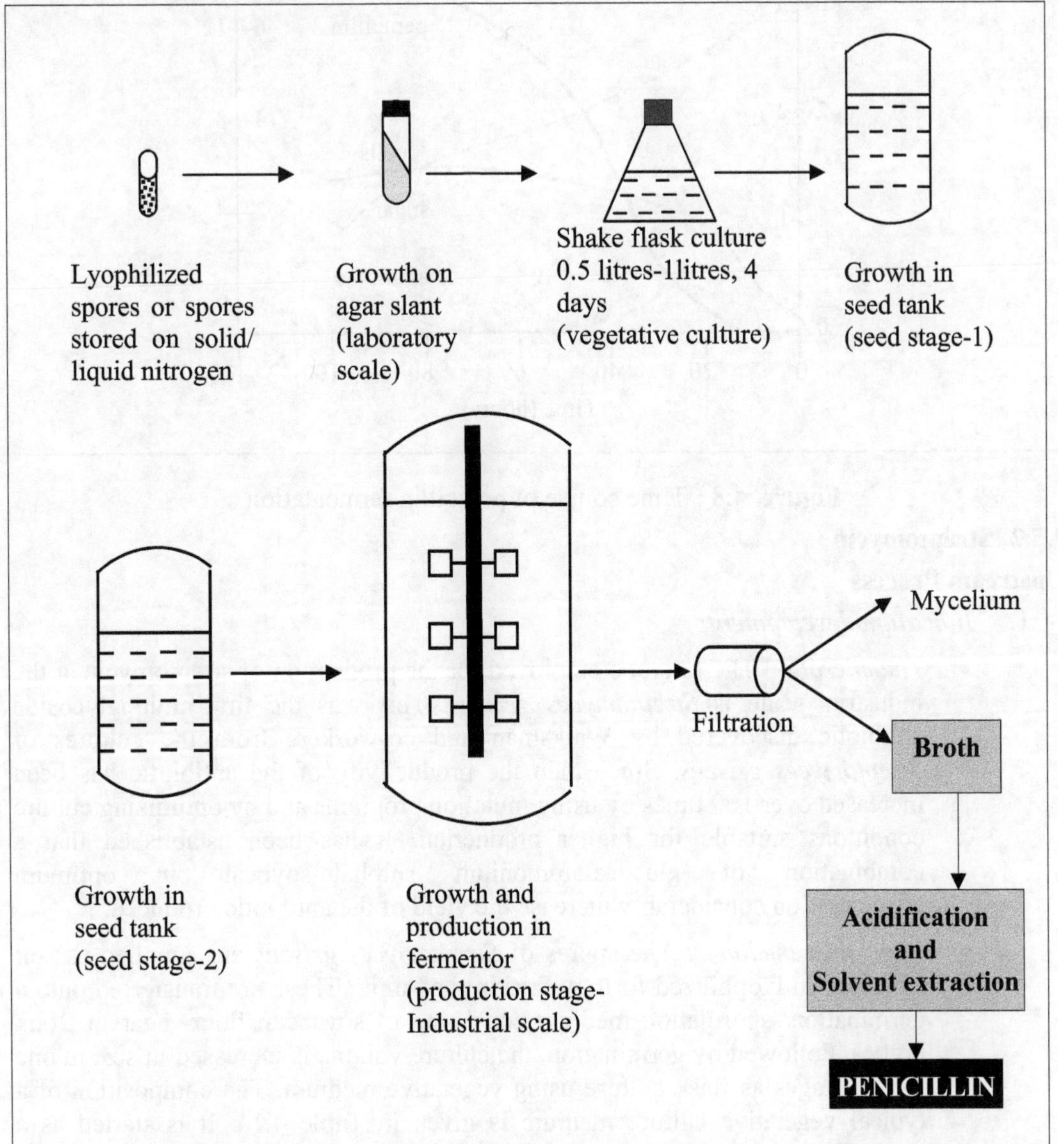

Figure. 4.7 : Flowchart for the fermentative production of penicillin.

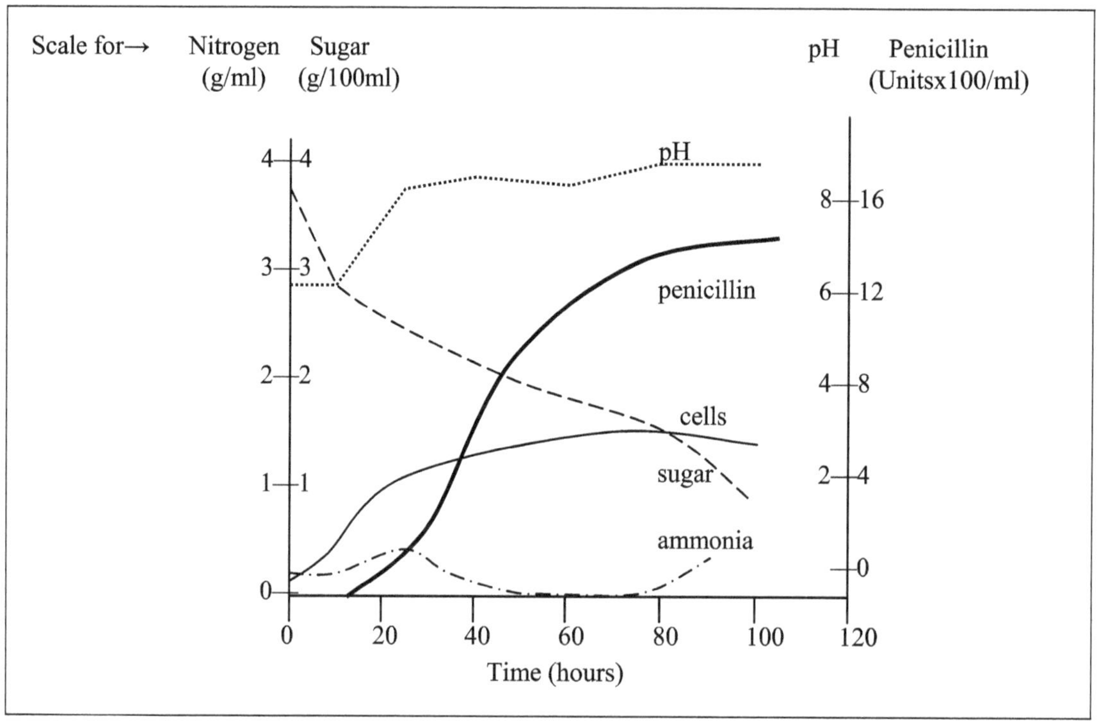

Figure. 4.8 : Time course of penicillin fermentation.

4.5.2 Streptomycin

Upstream Process

1. ***Inoculum development:***

   - *The microbe :* The microbe employed for the production of streptomycin at the industrial scale is *Streptomyces griseus.* This was the first aminoglycoside antibiotic discovered by Walksman and co-workers from the cultures of *Streptomyces griseus.* Since then the productivity of the antibiotic has been increased over 100 times by using mutation programs and by optimizing culture conditions suitable for higher production. It has been established that a combination of glucose-ammonium sulphate-soybean in optimum concentration considerably increase the yield of the antibiotic produced.

   - *Vegetative medium :* The spores of *Streptomyces griseus* are supplied as soil stocks or in lyophilized form with skimmed milk. These are transferred onto a germination/ sporulation medium consisting of soyabean flour- agar in Roux bottles. Followed by germination, the culture volume is increased in size in one to two stages as flask culture using vegetative medium. The composition of a typical vegetative culture medium is given in Table 4.14. It is started as a shake-flask culture which gradually increases in volume. Aeration and agitation of the culture are must.

Table 4.14 Vegetative culture medium for *Streptomyces griseus*.

Ingredient	Concentration
Glucose	4%
Dextrin	0.2%
Soyabean flour	3%
Ammonium sulphate	0.55%
Calcium carbonate	0.65%
Sodium chloride	0.2%
Potassium dihydrogen phosphate	0.0005%

For vegetative culture-

- incubation temperature is 26°C

- culture is to be agitated at an rpm of 250 and

- time period for incubation is 48hours.

2. ***Preparation of culture medium:*** There are numerous culture media that have been employed for the production of streptomycin. Two of these are given in Table 4.15.

Table 4.15 Production culture medium for streptomycin.

Ingredient	Hockenhul medium	Singh's medium
Glucose	2.5%	6%
Soyabean meal	4%	3%
Distiller's dried solubles	0.5%	
Cornsteep liquour	-	0.4%
Ammonium sulfate	-	0.9%
Calcium carbonate	-	0.05%
Sodium chloride	0.25%	0.25%
KH_2PO_4	-	0.0025%
Soyabean oil	-	0.7%
Distilled water	qs	qs

The pH of both the production media before sterilization is to be maintained at 7.3 to 7.5.

Additionally, there are four substances that stimulate and/ or assist in increasing the productivity of streptomycin. These include: barbital, factor A, myoinositol and scelerine.

3. *Sterilization:* Medium is sterilized by autoclaving (i.e. steam under pressure) at 120°C for 90 minutes. Note that glucose is sterilized separately (since it undergoes Millard reaction) and then mixed with the entire medium.

Fermentation

4. *Process control parameters/ characteristics :* The following are notable points concerning the production of streptomycin:

- Production is carried out by submerged culture method
- Production volume is $100m^3$-$200m^3$
- Aerobic process; aeration rate is $0.25m^3/h$
- Agitation is achieved by employing two impellers; at an rpm of 600
- During the process pH rises from 6.5-7,5-8.0
- Temperature should be controlled between 25°C-30°C
- Time period for completion i.e. when maximum yield is obtained is 5-7 days
- The production of streptomycin is phasic: First phase or growth phase or tropohophase- involves the growth of mycelium accompanied with utilization of protein content and release of ammonia; second phase or production phase or idiophase – involves the production of antibiotic accompanied with utilization of glucose; third phase- or death phase-involves cessation of the production brought about by the complete exhaustion of glucose in medium and cell lysis.
- Streptomycin is produced in a concentration of more than 1.2g/litre

Downstream process

5. *Recovery and purification :* There are several patented methods for the recovery of streptomycin from the fermentation broth. One such industrial process involves fixation of streptomycin on cationic exchange resins followed by elution using mineral acids. Steps involved in such a recovery process include:

- Filtration for removal of mycelium from broth
- Pretreatment of the filtered broth involving acidification, filtration and neutralization
- Fixation of streptomycin in the broth to a cation exchange resin by allowing the broth to run through a column containing Amberlite IRC-50.
- Elution in solutions and passage through different columns for removal of impurities
- Decolorization with carbon
- Elution with diluted hydrochloric acid
- Concentration under vacuum almost to dryness
- Dissolution in methanol and filtration

- Precipitation and washing using acetone
- Dissolution in methanol for preparation as streptomycin chloride.

4.5.3 Riboflavin

Upstream Process

1. *Inoculum development :*

- *The microbe :* There are a number of microbes that can be utilized for the production of riboflavin. These include *Clostridium acetobutylicum* and other bacterial species; *Candida flareri* and other species of yeast; *Bacillus subtilis* (recombinant bacteria) and yeast like microbes (ascomycetes) such as *Ashbya gossypi* and *Eremothecium ashbyii*. It is *Ashbya gossypii* that is most commonly employed for the production of riboflavin at the industrial scale. The principal reason as to why *Ashbya gossypii* is employed for production is due to its greater stability for producing riboflavin as compared to *Eremethecium asbyii* which has a greater tendency to degenerate to a form devoid of riboflavin-producing capacity. For enhancing the productivity, the strain has been developed by mutation programs and optimization of culture conditions.

- *Vegetative medium :* To initiate the fermentation process, *Ashbya gossypii* is transferred on a sporulation medium as given in Table 4.16. The culture volume is then increased, first in flask culture and then in a seed tank using the media shown in Table 4.17.

Table 4.16 Stock culture medium composition for *Ashbya gossypii*.

Ingredient	Concentration
Peptone	0.5%
Yeast extract	0.3%
Malt extract	0.3%
Glucose	1.0%
Agar	2.0%
Distilled water	qs

The incubation temperature for the stock culture is 27°C-30°C.

Table 4.17 Seed culture medium composition for *Ashbya gossypii*.

Ingredient	Concentration for Seed culture stage I	Concentration for Seed culture stage II
Glucose	2.0%	2.0%
Peptone	0.5%	-
Corn steep liquor	1.0%	1.0%
Animal stick liquor	-	0.5%
Distilled water	qs	qs

The pH of both the media is to be adjusted to 6.5 and the incubation temperature is to be maintained between 26°C-30°C

2. ***Preparation of production culture medium:*** A composition of a typical production medium for the production of riboflavin is shown in Table 4.18.

Table 4.18 Production culture medium composition for *A.gossypii.*

Ingredient	Concentration
Glucose	2.0%
Corn steep liquor	1.8%-2.1%
Animal stick liquor	1.0%
Antifoaming agent	qs
Distilled water	qs

The pH of the production medium is to be adjusted to 4.5.

3. ***Sterilization :*** is achieved by employing steam under pressure at a temperature of 135°C for 5 minutes.

Fermentation

4. ***Process control parameters/ characteristics:***

- Production is carried out by submerged culture method
- Production volume is 200 gallons-300 gallons
- Aerobic process; aeration rate is 0.25 vol. of air/ vol. of medium/ minute
- A high level of aeration should be avoided; since a high level would inhibit mycelial production and thereby the yield.
- The pH during the process is kept between 6-7.5
- Temperature should be controlled between 28°C-30°C
- Time period for completion i.e. when maximum yield is obtained is 4-5 days
- The production of ribolflavin is phasic: first phase or growth phase involves growth of microbe accompanied with utilization of glucose and accumulation of pyruvic acid (lowering of pH); second phase or production phase involves production of riboflavin (held inside the cell as FAD and FMN) accompanied with decrease in pyruvic acid levels and increase in ammonia levels (increase in pH); and third phase or the death phase involves cell autolysis and release of the vitamin in the broth.
- A yield of 4-2g/l is obtainable

Downstream process

5. ***Recovery and purification :*** If it is desired to use the product as feed supplement, then the broth at a pH of 4.5 is concentrated to syrupy consistency by evaporation using drum dryer. If a purified product is desired then there are numerous ways (all of which are patented) to recover the product. Some of these include:

- *Solvent extraction :* using butanol followed by extraction in petroleum ether/acetone
- *Precipitation :* using a reducing agent such as titanium chloride, followed by adsorption on diatomaceous earth

- *Adsorption :* on silica gel etc. followed by elution with aldehyde/ ketone/ alcoholic solution of an organic base
- *Microbial :* using reducing bacteria such as *Streptococcus faecalis.*

One of the methods for the isolation of riboflavin is illustrated schematically in Figure 4.9.

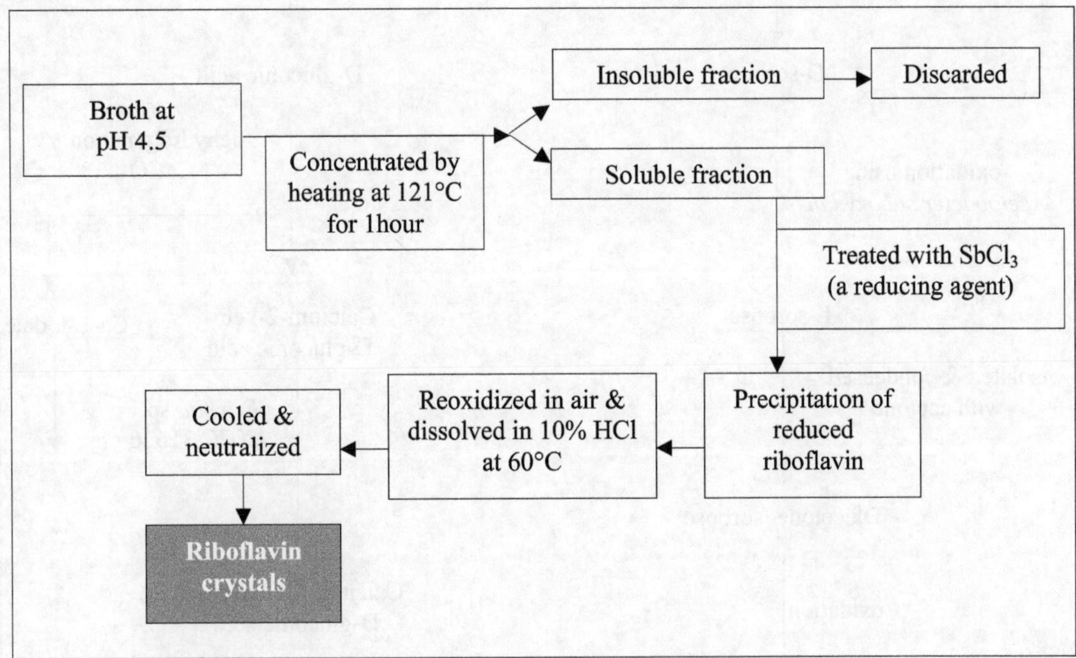

Figure. 4.9 : Recovery of riboflavin.

4.5.4 Ascorbic acid (Vitamin C)

L-Ascorbic acid is produced industrially using either crystalline glucose or hydrolysate of starch. There are two different processes/ pathways for the industrial production of Vitamin C as depicted in Figure 4.10. Both the processes involve the fermentative production of 2-keto-L-gluconic acid as the intermediate in the production of Vitamin C. First the fermentative process involving the use of *Acetrobacter suboxydans* shall be considered followed by the process involving the use of *Erwinia* and *Cornybacterium* sps.

4.5.4.1   Sorbose fermentation using Acetrobacter suboxydans

Upstream Process

1. *Inoculum development:*

- *The microbe :* The production of L-ascorbic acid starting with the reduction of glucose to sorbitol followed by fermentative conversion to sorbose employs *Acetobacter suboxydans*. It is to be noted that since *Acetrobacter suboxydans* is sensitive to nickel, strains have been developed that can tolerate high levels of the ion present in the fermentation broth.

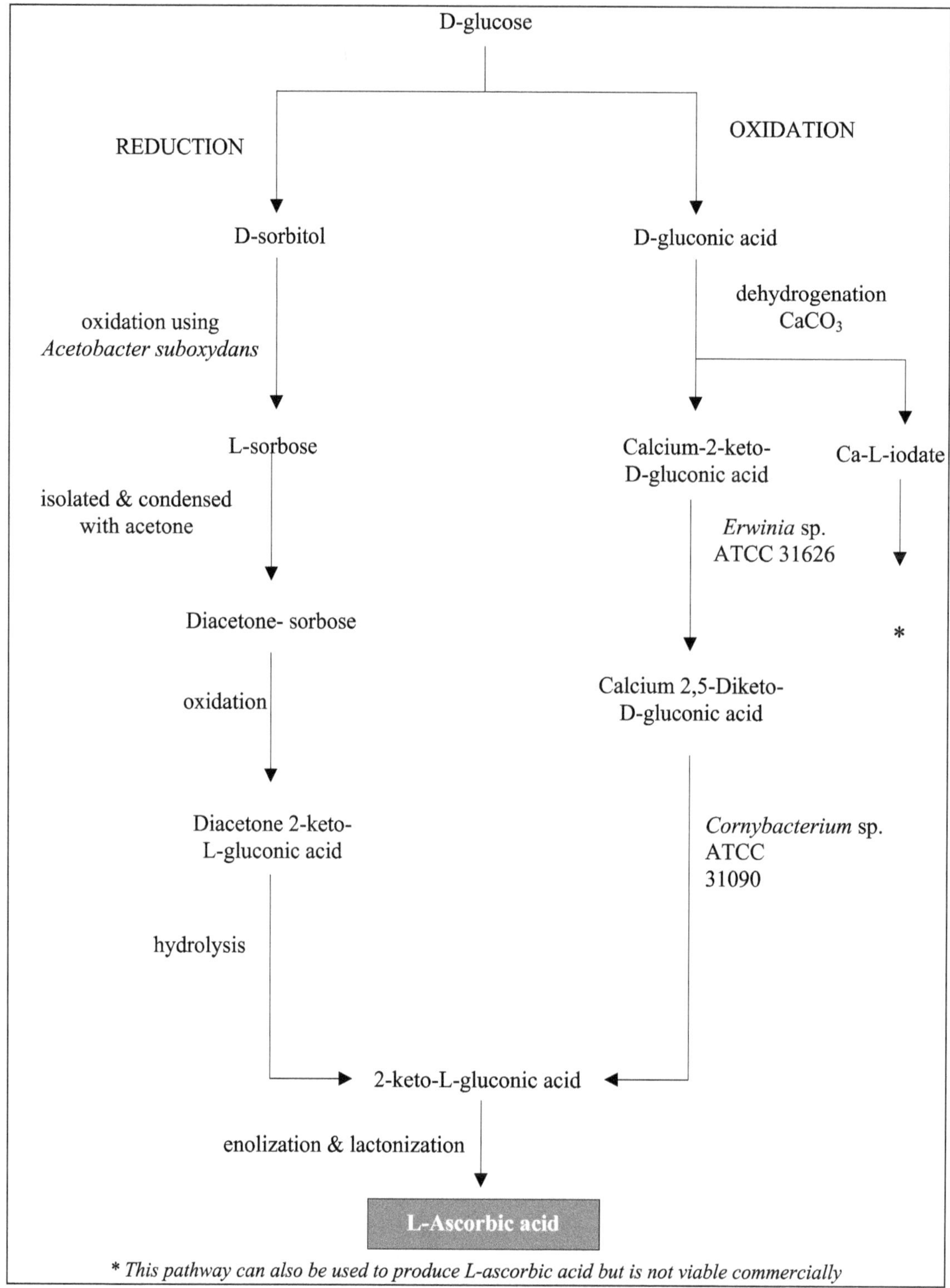

Figure. 4.10 Two pathways for the industrial production of L-ascorbic acid.

Vegetative medium : The composition of the inoculum culture medium for *Acetrobacter suboxydans* is presented in Table 4.19.

Table 4.19 Inoculum culture medium composition for *Acetrobacter suboxydans*.

Ingredient	Concentration
Sorbitol	10.0%
Yeast-extract	0.5%
Glucose	1.0%
Calcium carbonate	3.1%
Pantothenic acid	-
Distilled water	qs

2. ***Preparation of production culture medium :*** A composition of a typical production culture medium for the fermentative conversion of sorbitol to sorbose for production of L-ascorbic acid is shown in Table 4.20.

Table 4.20 Production culture medium composition for *A. suboxydans*.

Ingredient	Concentration
Sorbitol	20.0%
Corn steep liquor	0.1%-0.3%
Ammonium phosphate	3%-5%
Soyabean oil/ Octadecanol	0.1%
Distilled water	qs

3. ***Sterilization:*** is achieved by employing steam under pressure.

Fermentation

4. ***Process control parameters/ characteristics:***

    - Production is carried out by submerged culture method. Earlier, fermentations were based on surface method and rotating drum method.
    - Aerobic process; high aeration is required.
    - Time period for completion i.e. when maximum yield of sorbose is obtained is determined using a refractometer.

Downstream process

5. ***Recovery and purification:***

    - Filtration/ centrifugation to collect clear liquid (sugar sloution)
    - Deionization of filtrate (improves recovery)
    - Concentration under reduced pressure at a temperature below 50°C
    - Crystallization at a pH of 3.
    - Yield of sorbose, converted from sorbitol and recovered, is 87%.

4.5.4.2   2, 5-diketo-D-gluconic acid pathway using *Erwinia* and *Cornybacterium* sps

Upstream Process

1. ***Inoculum development :***

   - *The microbe :* The production of L-ascorbic acid starting with the oxidation of glucose to D-gluconic acid and then to 2-keto-D-gluconic acid followed by fermentative conversion to 2,5-diketo-D-gluconic acid employs *Erwinia* sp. (ATCC 31626)

     In the second stage, using 2,5-diketo-D-gluconic acid as the raw material, *Cornybacterium* sp. (ATCC 31090) convert it into 2-keto-L-gluconic acid.

   - *Vegetative medium :* The composition of the vegetative culture medium for *Erwinia* sp. is presented in Table 4.21.

Table 4.21 Vegetative culture medium composition for *Erwinia* sp.

Ingredient
Glycerol
Yeast extract
Peptone
Potassium dihydrogen phosphate
Magnesium Sulphate
Agar
Distilled water

The incubation temperature for vegetative culture is 28°C and incubation time is one day. Further, for inoculum development, a laboratory culture and a seed culture are prepared and inoculated with the culture as developed from the above vegetative stage. The composition of both (laboratory and seed culture) development media is similar and is presented in Table 4.22.

Table 4.22 Inoculum culture medium composition for *Erwinia* sp.

Ingredient	Concentration
Glycerol	0.5%
Corn steep liquor	3.0%
Potassium dihydrogen phosphate	0.1%
Soyabeal oil	trace amount as antifoaming agent
Distilled water	qs

The incubation temperature for the inoculum culture is 28°C and the incubation time 20 + 10 hours. with aeration and agitation.

2. ***Preparation of production culture medium:*** A composition of a typical production medium for the fermentative production of 2,5-diketo-D-gluconic acid is shown in Table 4.23.

Table 4.23 Production culture medium composition for *Erwinia* sp.

Ingredient	Concentration
Glucose	5.8%
Corn steep liquor	1.1%
Ammonium acid phosphate	3%-5%
Calcium carbonate	18.4%
Soyabean oil/ Octadecanol	0.02%
Distilled water	qs

3.   *Sterilization :* is achieved by employing steam under pressure.

Fermentation

4.   *Process control parameters/ characteristics:*

- Production is carried out by submerged culture method.
- Aerobic process; requiring 1.5v/v of air at a pressure of 1.5 kg/cm^2
- Agitation is provided at an rpm of 160.
- Carried out at a temperature of 28°C
- Time period for completion i.e. when maximum yield of calcium 2,5-diketo-D gluconic acid is obtained is 26 hours.

***Cornybacterium* sps**

- Subsequently, calcium 2,5-diketo-D-gluconic acid is fermentatively converted into calcium 2-keto-L-gluconic acid by employing *Cornybacterium* sp., making it a dual fermentation.
- The compositions of vegetative culture, first & second seed culture and production media are given in Tables 4. 24, 4.25 and 4.26.

Table 4.24 Vegetative culture medium composition for *Cornybacterium* sp.

Ingredient	Concentration
Glucose	0.5%
Yeast extract	0.5%
Peptone	0.5%
Potassium dihydrogen phosphate	0.1%
Magnisium Sulfate	0.02%
Agar	2.0%
Distilled water	qs

For vegetative culture-

- Incubation temperature is 28°C and
- Time period for incubation is 40hours.

Table 4.25 Seed culture medium composition for *Cornybacterium* sp.

Ingredient	Concentration	
	First stage	**Second Stage**
Glucose	1.0%	1.0%
Yeast extract	0.5%	-
Corn steep liquor	-	2.0%
Peptone	0.5%	-
Sodium nitrite	0.1%	0.2%
Potassium dihydrogen phosphate	0.1%	0.1%
Magnisium Sulfate	0.02%	0.02%
Distilled water	qs	qs

For seed culture, first stage,-

- Incubation temperature is 28°C
- Culture is to be incubated in rotary shaker and
- Time period for incubation is 24hours.

For vegetative culture, second stage,-

- Incubation temperature is 28°C
- Culture is to be incubated in seed tank
- Culture is to be agitated at an rpm of 244 and
- Time period for incubation is 24hours.

Table 4.26 Production culture medium composition for *Cornybacterium* sp.

Ingredient	Concentration
Glucose	2.0%
Cornsteep liquor	3.0%
Sodium Nitrite	0.35%
Potassium dihydrogen phosphate	0.06%
Sodium Nitrite	0.1% (added after 16 hours)
Zinc sulphate	
Manganese chloride	
Thiamine HCl	trace amounts
Calcium D-pantothenate	
Distilled Water	qs

Process parameters:

- Production is carried out by submerged culture method.
- Aerobic process; requiring 0.25 v/v of air per minute.
- Agitation is provided at an rpm of 160.
- Carried out at a temperature of 28°C.
- After 18 hours, broth containing calcium 2, 5-diketo-D-gluconic acid from *Erwinia* culture is added.
- Time period for completion i.e. when maximum yield of calcium 2-keto-L-gluconic acid is obtained is 66 hours.

Downstream process

5. *Recovery and purification:*

   - Filtration/ centrifugation to collect clear liquid
   - Concentration
   - Treatment with sodium carbonate
   - Crystallization
   - Yield of 2-keto-L-gluconic acid is 90%.

4.5.5 L-Lysine

There are two methods used for the production of L-lysine: indirect or dual fermentation and direct fermentation. Both of these are considered are below.

4.5.5.1 Indirect or dual fermentation:

Upstream Process

1. *Inoculum development:*

   - *The microbe :* The fermentative production of lysine is an example of a dual fermentation i.e. two microbes are involved; first microbe produces a product which serves as the raw material for the second microbe. Thus, in the first part of the fermentation, an *Escherichia coli* auxotroph mutant (ATCC 12,408) is employed and in the second part *Aerobacter aerogenes* (ATCC 12,409) is employed.

     By definition, an auxotrophic mutant is a mutated microbe that has lost the ability to produce one or more enzymes of a biosynthetic pathway and hence requires the inclusion of the particular preformed metabolite(s) (which is/ are otherwise produced by the enzyme) in the medium for its growth. Such an auxotrophic mutant is useful in fermentations since it accumulates the intermediate compound(s) in the culture medium as it lacks the enzyme(s) to convert it further into its product(s).

     In this fermentation, the mutant auxotroph of *Escherichia coli* lacks the enzymes lysine decarboxylase and α, ε-diaminopimelic acid decarboxylase (Figure 4.11). Because of the latter metabolic block, it becomes necessary to add small amounts of lysine for the growth of the microbe; at the same time the advantage is that it allows for the accumulation of huge amounts of α, ε-diaminopimelic acid (DAP). Now this DAP (accumulated in large amonts) serves as the raw material for the second microbe: *Aerobacter aerogenes* which converts it into L-lysine (so that a high yield is obtained).

It is important to note that *Escherichia coli* should also possess a strong α, ε-diaminopimelic acid racemase activity and that both *Escherichia coli* and *Aerobacter aerogenes* should lack the ability to produce lysine decarboxylase. This enzyme is responsible for further converting lysine into cadaverine, which could be detrimental to the yield of lysine that can be obtained.

- *Vegetative medium :* The composition of the culture medium for *Escherichia coli* that is used as the inoculum for the production medium is presented in Table 4.27.

Table 4.27 Inoculum culture medium composition for *E.coli.*

Ingredient	Concentration
Glycerol	0.5%
Corn-steep liquor	0.5%
$(NH_4)_2HPO_4$	0.5%
Distilled water	qs

For inoculum culture-

- pH is to be adjusted to 7.5

- Incubation temperature is 28°C and

- Time period for incubation is 20hours.

2. ***Preparation of production culture medium :*** A composition of a typical production medium for the production of L-lysine is shown in Table 4.28.

Table 4.28 Production culture medium composition for *E. coli.*

Ingredient	Concentration
Glycerol	6%
Corn-steep liquor	4%
$(NH_4)_2HPO_4$	4%
Calcium carbonate	0.5%
Lysine	(supplied from corn-steep liquor)
Soyabean oil	trace amounts as antifoaming agent
Distilled water	qs

For production culture-

- pH is to be adjusted to neutral to alkaline.

3. ***Sterilization :*** is achieved by employing steam under pressure of 20 psi at a temperature of 121°C for 30 minutes.

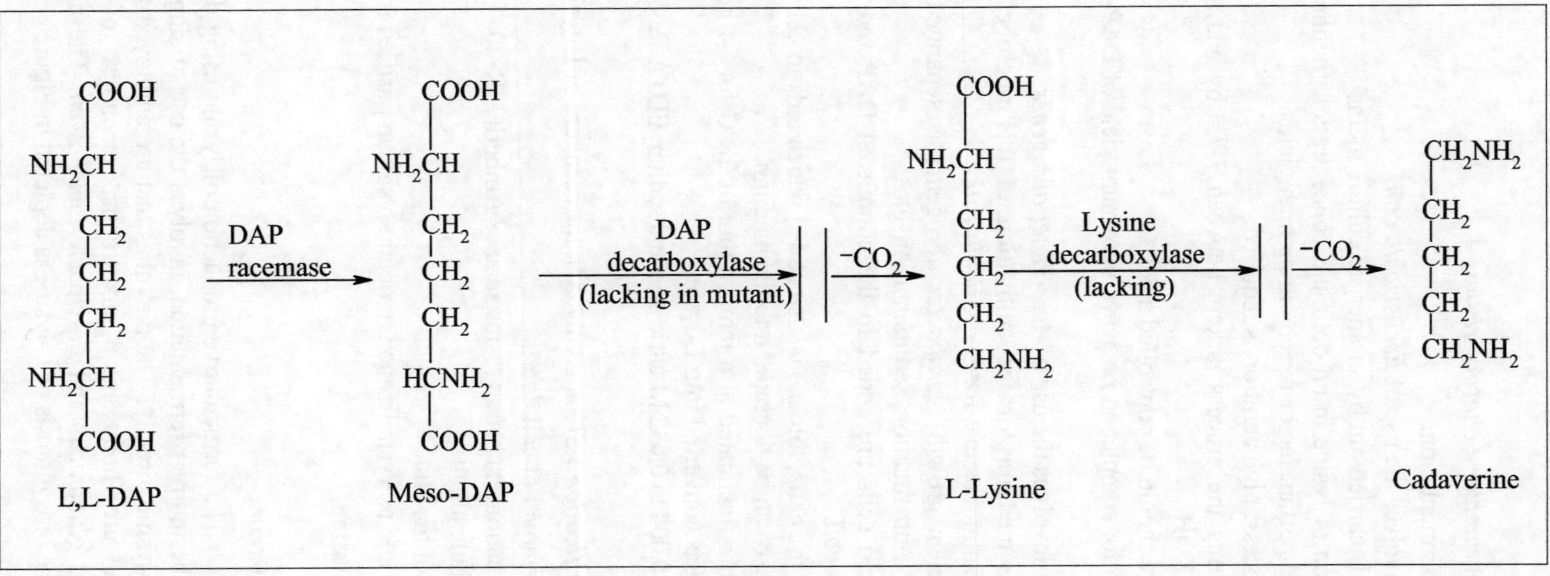

Figure. 4.11 : Depicting metabolic pathway leading to the biosynthesis of L-lysine.

Fermentation

4. *Process control parameters/ characteristics:*
   - It is a dual fermentation.
   - First part involves the use of *Escherichia coli*
   - Production is carried out by submerged culture method.
   - Incorporation of lysine in an optimum concentration (neither too high nor too low) is an important factor for obtaining high yields.
   - Aerobic process; high aeration is required.
   - The pH during the process is kept between 7-7.5 by addition of ammonium hydroxide/ KOH.
   - Temperature should be controlled at 28°C.
   - Time period for completion i.e. when maximum yield of DAP is obtained is three days.
   - Second part involves the use of *Aerobacter aerogenes.*
   - The culture medium for the cultivation of this microbe is similar to the inoculum medium as that for *Escherichia coli.*
   - After sufficient growth, *Aerobacter aerogenes* is separated from the growth medium by centrifugation, sedimentation, etc.
   - The collected cells are added to the completed DAP fermentation broth of *Escherichia coli.*
   - Along with the cells, toluene is also added which aids in cell lysis so that DAP decarboxylase can be liberated into the medium.
   - The ferment is incubated at a temperature of 28°C for 24 hours during which time DAP gets converted into L-lysine.
   - Addition of 0.004 to 0.032 M citric acid and sodium EDTA favour the conversion.

Downstream process

5. *Recovery and purification :* Steps involved in the recovery of L-lysine include:
   - Filtration for removal of cell debris
   - Adsorption on cation-exchange resins such as Amberlte IR-20
   - Elution with dilute alkali
   - Concentration of the elute
   - Crystallization and recrystallization to obtain lysine in purified form.

4.5.5.2   Direct fermentation

Upstream Process

1. *Inoculum development:*
   - *The microbe :* The fermentative production of lysine using direct fermentation i.e. unlike the indirect fermentation, involves the use of single microbe which produces L-lysine directly. The microbes that are employed for the production of L-lysine are homoserine auxotrophs; examples of which include: *Micrococcus glutamicus* and *Brevibacterium flavum*. The metabolic pathway leading to the biosynthesis of L-lysine is depicted in Figure 4.12.

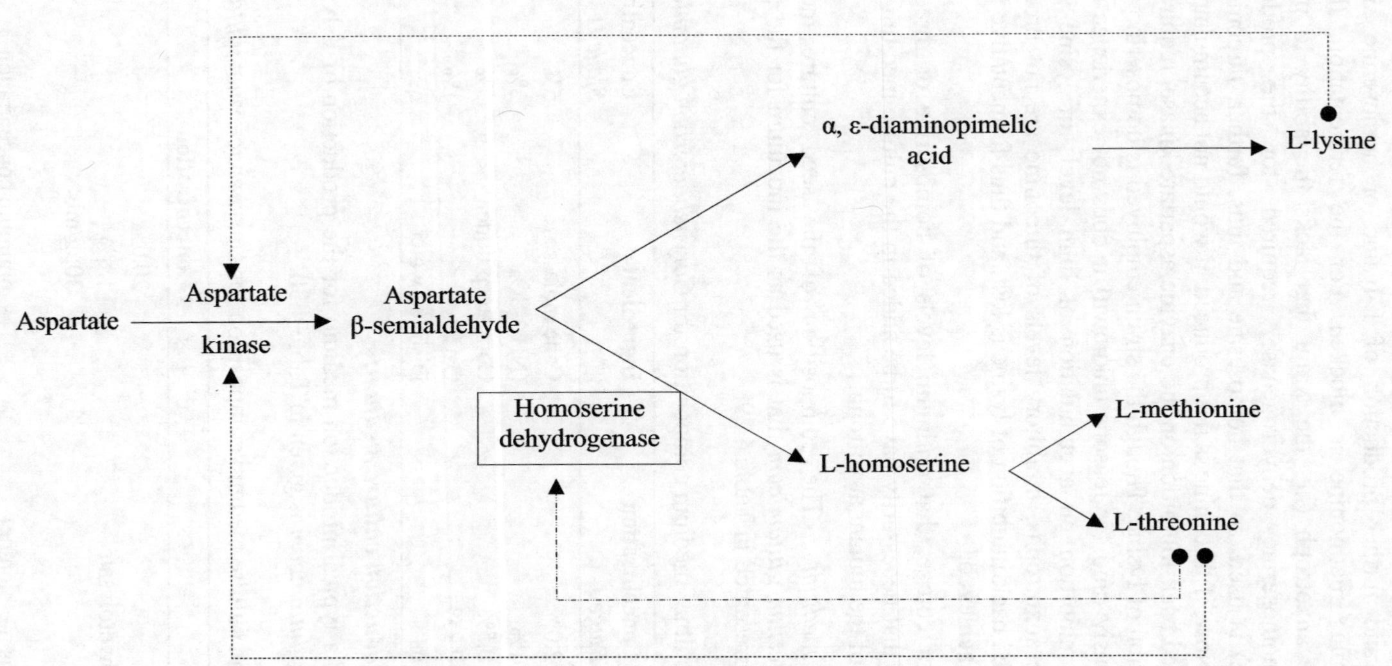

Figure. 4.12 Negative feed back control over the production of L-lysine and L-homoserine in the prototroph.

153

In the natural/ wild type/ prototroph strains, accumulation of L-lysine or L-threonine results in negative feed back control either through inhibition or repression and further production of L-lysine or homoserine and further threonine plus methionine is blocked. For the fermentation, the mutant homoserine auxotroph (i.e. the strain that lacks the ability to produce an enzyme, homoserine dehydrogenase, required for the production of homoserine) is used, so that homoserine and thus, further, threonine are not produced. Now, if threonine is not produced, it would not accumulate to exert a negative feed back inhibition on the enzyme aspartate kinase, required also for the production of lysine. Though the strain employed is homoserine auxotroph, advantageously (and for reasons unknown) it does not experience a negative feed back inhibition on accumulation of high levels of lysine. Using the homoserine auxotrophs, inhibitory levels of threonine are not produced and negative feed back inhibition of lysine is lost, and thus fermentative production of lysine is achieved.

It is, but of course, that optimum levels of homoserine or threonine plus methionine that necessarily have to be added in the culture medium to support the growth of the mutant auxotroph.

- *Vegetative medium :* The composition of the seed culture medium for *Corynebacterium glutamicum* that is used as the inoculum for the production medium is presented in Table 4.29.

Table 4.29 Seed culture medium composition for *Corynebacterium glutamicum*.

Ingredient	Concentration *Stage I*	Ingredient	Concentration *Stage II*
Glucose	2.0%	Cane molasses	5%
Peptone	1%	$(NH_4)_2SO_4$	2%
Meat extract	0.5%	Corn steep liquor	5%
NaCl	0.25%	$CaCO_3$	1%
Distilled water	qs	Distilled water	qs

(ii)   *Preparation of production culture medium:*

A composition of a typical production medium for the production of L-lysine using *Corynebacterium glutamicum* is given in Table 4.30.

Table 4.30 Production culture medium composition for *Corynebacterium glutamicum*.

Ingridient	Concentration
Cane molasses	20.0%
Soyabean meal hydrolysate	1.8%
Biotin	30µg/ml<
Homoserine	
(or threonine plus methionine)	optimum concentration
Distilled water	qs

3. ***Sterilization :*** is achieved by employing steam under pressure.

Fermentation

4. ***Process control parameters/ characteristics:***
   - Production is carried out by submerged culture method.
   - Incorporation of biotin in an optimum concentration is an important factor so that production of L-glutamic acid does not occur.
   - Incorporation of L-homoserine (or threonine plus methionine) in an optimum concentration (neither too high nor too low) is an important factor for obtaining high yields. Since a high level would exert negative feed back inhibition while too low a concentration would not allow growth of the microbe, both of which shall be detrimental for the production and yield of L-Lysine.
   - Temperature should be controlled at 28°C.
   - Time period for completion i.e. when maximum yield is obtained is 6 hours.
   - Maximum yield obtained is 44g/L.
   - Production using mutant homoserine auxotroph of *Brevibacterium Flavum* employs a 60%w/v solution of acetic as the carbon source.
   - Fermentation is carried out at a temperature of 33°C and the maximum yield obtained is 75g/L of L-lysine.

4.5.6 L-Glutamic acid

This amino acid like L-lysine is produced by direct (single stage) or indirect (two stage) fermentation process. Both of these processes are considered below.

4.5.6.1   Direct/ One-stage fermentation:

Upstream Process

1. ***Inoculum development:***
   - *The microbe :* The fermentative production of L-glutamic acid using direct fermentation i.e. unlike the indirect fermentation, involves the use of single microbe which produces L-glutamic acid directly. Examples of such microbes that produce L-glutamic acid are given in Table 4.31. The metabolic pathway leading to the biosynthesis of L- glutamic acid is depicted in Figure 4.13.

Table 4.31 Examples of some microbes that produce L-glutamic acid

Arthrobacter globformis	*Brevibacterium flavum*
Bacillus megaterium cereus	*Cornybacterium herculi*
Brevibacterium aminogenes	*Micrococcus glutamicus*

The microbial strain that is commonly employed for the fermentative production of L-glutamic acid commercially is *Micrococcus glutamicus*, No. 541.

2. ***Sterilization :*** is achieved by employing steam under pressure.

Fermentation

3. *Process control parameters/ characteristics:*

   - Production is carried out by submerged culture method.
   - Aerobic process; high aeration is required.
   - The pH during the process is kept between 6 and 8.
   - Temperature should be controlled at 30°C.
   - The process involves intermittent addition of urea, ammonia or ammonium salts which serve as source of ammonium ions.
   - Permeability barrier to glutamic acid secretion is overcome by one of the following:
     use of optimum biotin concentration, addition of penicillin or addition of surfactants
   - Time period for completion i.e. when maximum yield is obtained is little less than two days
   - The obtainable yield is 30g/L.

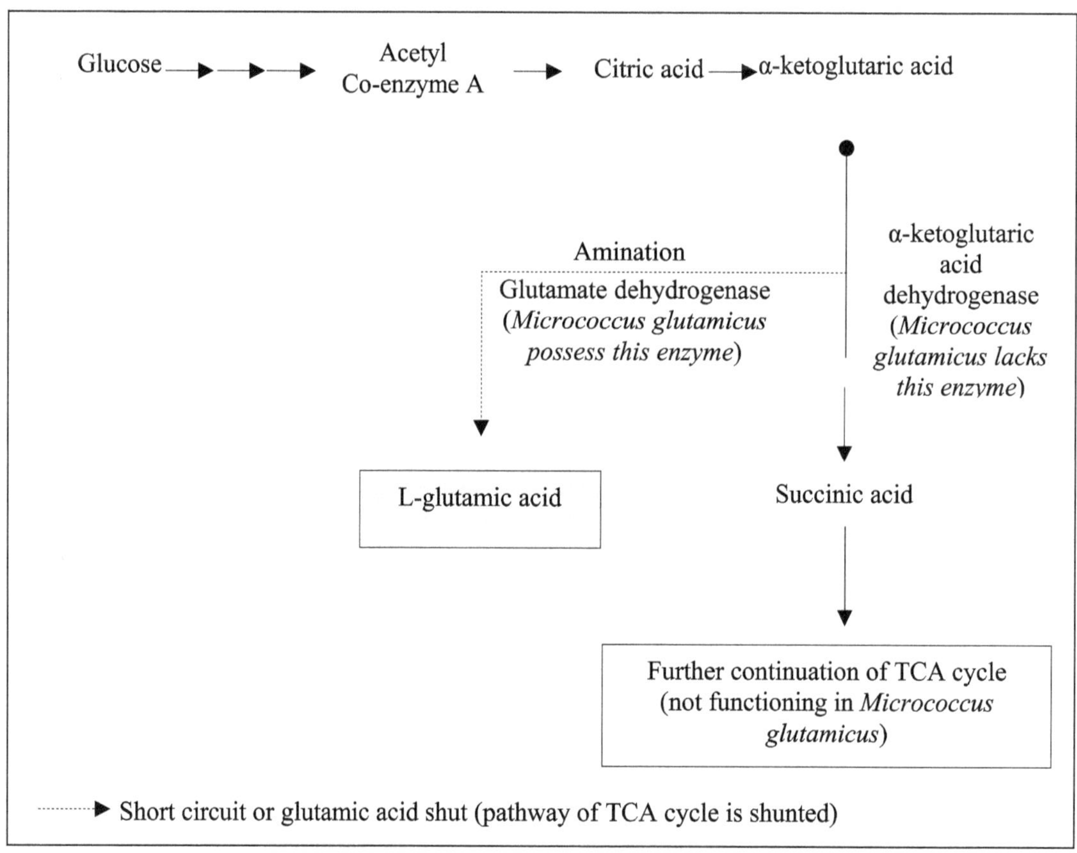

Figure. 4.13 The metabolic pathway leading to the biosynthesis of L- glutamic acid.

Downstream process

4. *Recovery and purification:*

The recovery of L-glutamic acid is achieved in the following manner:

- Filtration for removal of microbial cell mass.
- Concentration of the filtered broth
- Adsorption on ion-exchange resins
- Elution
- Concentration of the elute
- Crystallization and recrystallization to obtain L-glutamic acid in purified form.

4.5.6.2 Indirect/ two stage fermentation

L-Glutamic acid is also produced using a fermentation process involving two stages. In the first stage, the intermediate, α-ketoglutaric acid is produced by one microbe; then in the second stage, it is converted to L-glutamic employing another microbe. Microbes employed for the first and second stages are presented in Table 4.32.

Table 4.32 Examples of some microbes employed in two stage process for production of L-glutamic acid

Microbes producing α-ketoglutaric acid in the first stage	Genera of microbes converting α-ketoglutaric acid to L-glutamic acid in the second stage
Bacillus magatherium	*Aerobacter*
Bacterium α-ketoglutaricum	*Bacillus*
Escherichia coli (no.3 & 7)	*Erwinia*
Escherichia freundii (no.5, 9, 12)	*Escherichia*
Kluyvera citrophila(no 11, 84C)	*Hansenula*
Pseudomonas fluroscens	*Mycotorula*
Serratia marcescens	*Pseudomonas*
	Serratia
	Xanthomonas

4.5.7 α-Amylases

α-Amylases are produced both by fungi and bacteria. Table 4.33, lists the microbes employed for the production of α- amylases.

Therefore, depending upon the microbe used for production, α- amylases are referred to as either fungal or bacterial α- amylases. Refer Table 3.5 for a comparison of the two types of amylases. We shall first consider the production of fungal amylases following by production of bacterial amylases.

Table 4.33 Microbes employed for the production of α- amylases

Fungi	Bacteria	
Aspergillus oryzae	*Bacillus subtilis*	*Cl. acetobutylicum*
Aspergillus niger	*B. diastaticus*	*Sarcina* sps.
	B. polymyxa	*Actinomyces* sps.
	B. macerans	*Phytomonas* sps.

4.5.7.1   Fungal amylases

Upstream Process

1. ***Inoculum development***

   - *The microbe : Aspergillus oryzae is the strain used when solid substrate cultivation technique is used for the production of the enzyme. Aspergillus niger is the strain used when submerged cultivation technique is used for the production of the enzyme.*

     It is interesting to note that the substrate/ production medium is inoculated with *Aspergillus oryzae* spores in moist and dry forms. Figure 4.14, depicts the scheme for developing moist mould-spore preparation to be used for substrate inoculation.

     Spores as supplied in the dry form are also used to inoculate the substrate in a concentration as low as 0.04% of production medium.

2. ***Preparation of production culture medium:*** The medium, for the production of fungal amylases using solid substrate fermentation, which is universally used, is wheat bran. Alternatively, other substrates such as rice bran, soybean or sweet potato flakes or suitably size-reduced grains can also be used.

   For production and growth, it is further only required that the substarte be moistened with water or dilute acid. The amount of water needed for moistening the substarte is in the range of 40%-70%.

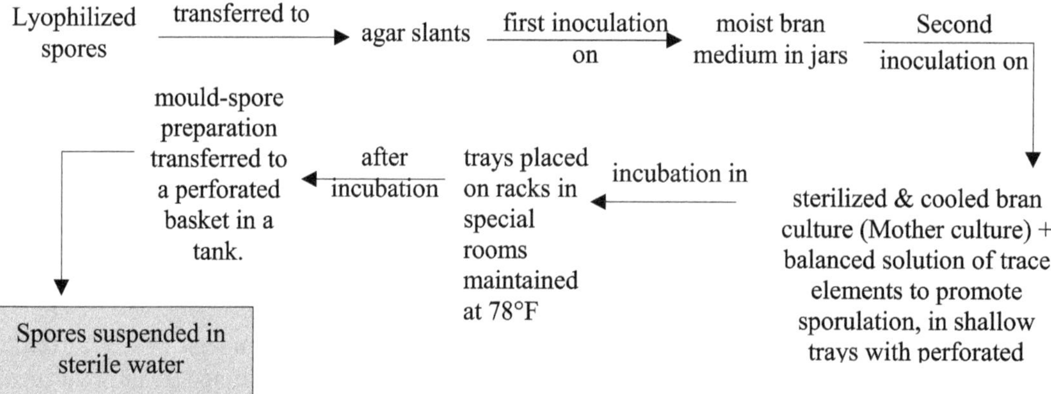

Figure. 4.14 Development of moist spores for inoculation of production medium.

3.   *Sterilization :* is achieved by one of the following ways:

- introduction of live steam into the bran, if water has been used to moisten the bran
- heating at 95°C for 15-30 minutes, if dilute acid has been used to moisten the bran use of bactericide such as formaldehyde or beta-propio-lactone (BPL)

Fermentation

4.   *Process control parameters/ characteristics:* The methods for cultivation of moulds used for the production of amylase include the following:

- solid substrate culture method: under this method, there are three techniques:
    - tray process/ thin layer process/ cabinet method
    - deep bed process and
    - rotary drum process.
- submerged culture method in liquid medium
    - Tray process:
- The process is mechanized and automated and is depicted in Figure 4.15.
- The moistened bran along with the inoculum (moist and/or dry spores) is uniformly distributed in trays.
- The height of substrate layers in trays is 2-4 cm.
- The trays may be either rectangular or circular in shape and may be made either of metal or wood.
- The trays are loaded in a cabinet or truck and conveyed to the culturing tunnel/ special incubators.
- Temperature is controlled at 37°F and humidified air is circulated during incubation.
- Time period for incubation is 24-30 days.
- At the end of this, the cabinet/ truck carrying the trays moves over to the drying tunnel.
- Deep bed process:
- It is an automated process.
- The depth of substrate layer (bed) ranges from as low as 0.6m to as high as 1.5 -1.8m.
- The beds of substrate travel automatically through the culturing tunnels by means of conveyer belts.

It is to be noted that when solid substrate cultivation techniques are used, control of moisture, temperature, aeration and pH becomes a difficult task. The solution to this problem is solved by selection of strains that grow slowly, coupled with the characteristic of producing a high yield.

Downstream process

5.   *Recovery and purification:*

- The mass is dried at a temperature not exceeding 50°C such that the moisture content is reduced to 8%.
- Dried material is ground and bagged.

Alternatively, amylase in pure form can be obtained by:

- Extraction in water from the production medium
- Precipitation by addition of alcohol
- Drying at a temperature below 55°C.

Submerged technique :

- *Microbe* : The strain suited for production of amylase using this cultivation technique is *Aspergillus niger*
- Raw materials are mixed in a tank, heated, cooked and sterilized.
- The medium is then cooled to a temperature of 37°F.
- Sterile media is transferred to a fermentor and inoculated.
- Proper control of aeration, agitation and pH is needed.
- *Recovery*: This involves filtration to remove the mycelium.
- Enzyme is separated from the liquor by either salt or solvent precipitation.
- The precipitated enzyme is further processed/ crystallized to achieve higher purity.

4.5.7.2   Bacterial amylases

A number of workers have reported/ worked on the production of amylase by bacteria. Procedures for the production of bacterial amylase by three such workers are presented in Table 4.34.

4.5.8 Proteases

Proteases are produced both by fungi and bacteria. Table 4.35 lists the microbes employed for the production of proteases.

Table 4.35 Microbes producing proteases.

Fungi	Bacteria
Aspergillus niger	*Bacillus*
Aspergillus oryzae	*Clostridium*
Aspergillus flavus	*Proteus*
Amylomyces rouxii	*Pseudomonas*
Mucor delemar	*Serratia*
Penicillium roquefortii	

Fungal and bacterial proteases are produced under fermentation conditions similar to those employed for the production of amylases. Except that a *Bacillus subtilis* starin producing high yields of protease (and not amylase) is selected and the medium contains a high proportion of carbohydrate to stimulate protease production and depresses amylase production.

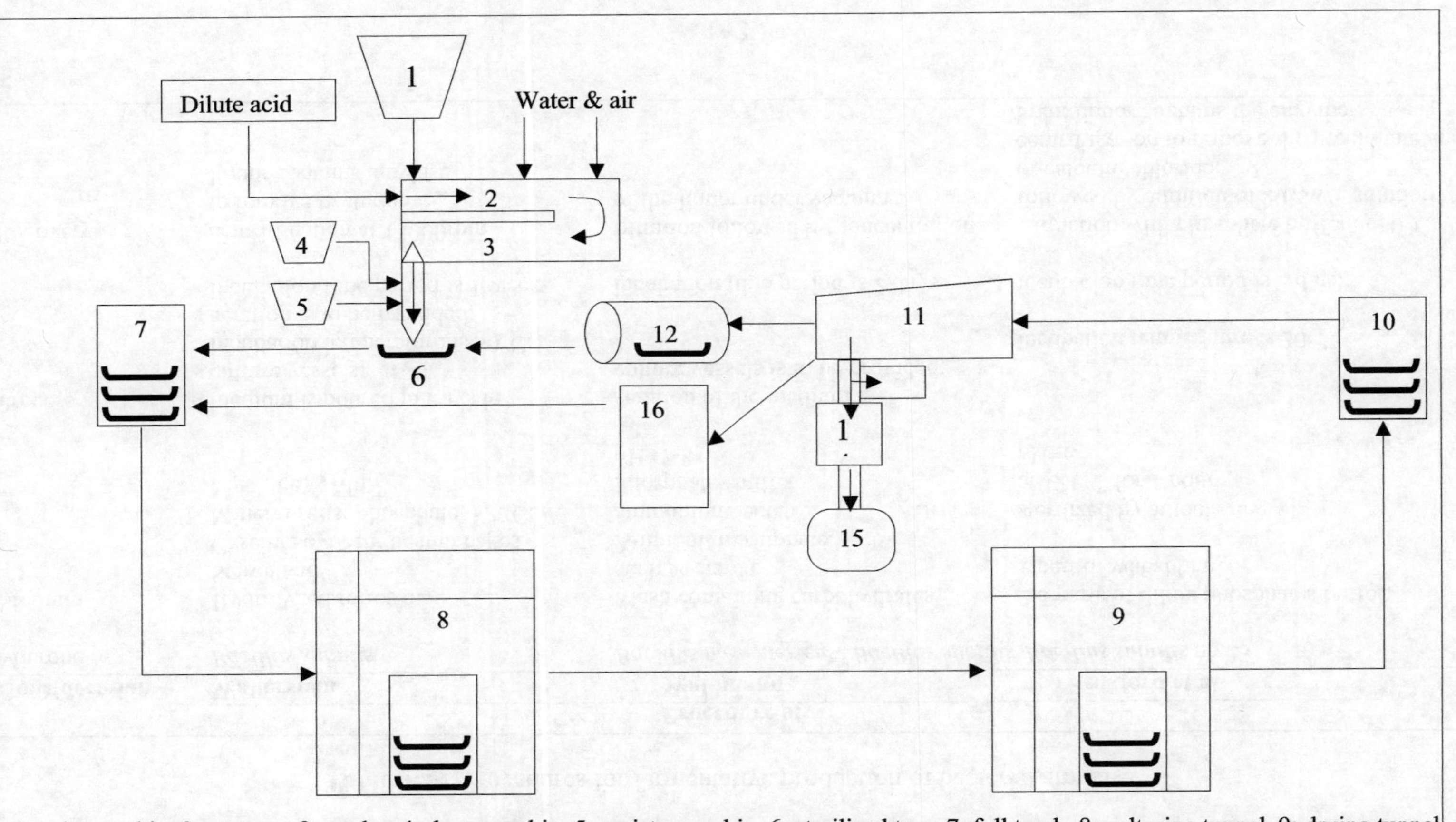

1: substrate bin. 2: steamer. 3: cooler. 4: dry spore bin. 5: moist spore bin. 6: sterilized tray. 7: full truck. 8: culturing tunnel. 9: drying tunnel. 10: full truck (amylase produced). 11: discharge bin. 12: sterilizer. 13: extractor. 14: grinder. 15: bagger. 16: empty truck.

Figure. 4.15 Flow sheet for the production of fungal amylase using tray process.

Table 4.34 Procedures for fermentative production of bacterial amylase.

Consideration	Procedure of		
	Wallerstein	Waldmann	Beckford et al
Microbe	*Bacillus subtilis*	*Bacillus mesentericus; Bacillus subtilis*	*Bacillus subtilis* no.23
Medium	Hydrolyzed forms of: Starch and Casein/ soybean/ peanut cakes Mineral salts: phosphate, K^+, Ca^{2+}, Mg^{2+}, Fe^{2+}, Mn^{2+}	Mash containing carbohydrates such as starch; Ammonium sulphate, ammonium lactate; Phosphates buffer pH=7-8	2.5 parts of dilute phosphates buffer, 1 part of wheat bran sterilized by autoclaving at 121°C for 1 hour pH=6
Process	medium is poured in trays in culture vessels; incubation temperature is 30°C; aeration is to be provided; incubation time period is 1 week	aeration of the medium in culture vessels is to be provided; incubation time period is 2 days	incubation temperature is 30°C; incubation time period is 2 days
Recovery	centrifugation at 14000rpm to remove bacteria; liquor contains amylase	filtration followed by concentration of the liquor under vacuum	extraction with phosphate buffer at 30°C; followed by addition of 20%w/v solution of calcium chloride; centrifugation to remove the precipitate and bran; liquor contains the enzyme

5 # PLANT CULTURE

CHAPTER CONTENTS AT A GLANCE

5.1   Introduction

Plant culture refers to the *in vitro* proliferation of parts of plants. It may involve the culturing of plant protoplasts, cells, tissues or organs obtained from natural strains or improved strains produced by way of selection, mutation or genetic engineering. The plant part, called as explant, can only be cultured *in vitro* when all the required nutrients, in the right quantities and the right conditions are provided in the artificially prepared plant culture medium. The medium may be liquid or solid. It is essential to carry out any of the manipulations in plant culture using aseptic technique, since the contaminating microbes (bacteria, fungus or virus) entering into the culture will outgrow, growing at a faster rate than the desired plant cells/tissue.

The explant transferred in culture results in the production of a mass of cells called callus which can be disintegrated to produce suspension cell culture or it can be induced to form organs (organogenesis), or to generate plantlets.

5.2   Scope

Plant culture has wide ranging applications in many areas which include:

1.   plant cloning
2.   plant genetic engineering
3.   production of (animal) protein in plants
4.   production of secondary plant metabolites

A list of compounds reported from plant culture is presented in Table 5.1.

Table 5.1 Substances reported from plant cultures.

Alkaloids	Enzyme inhibitors	Opiates
Allergins	Flavanoids, flavones	Organic acids
Anthraquinones	Flavours	Peptides
Antileukaemic agents	Furacoumarins	Perfumes
Antitumour agents	Hormones	Phenols
Antiviral agents	Insecticides	Pigments
Aromas	Interferons	Plant growth regulators
Benzoquinones	Latex	Proteins
Carbohydrates	Lipids	Steroids and derivatives
Cardiac glycosides	Naphthoquinones	Sugars
Chalcones	Nucleic acids	Tannins
Diathrenes	Nucleotides	Terpenes, Terpenoids
Enzymes	Oils	Vitamins

5.3   History

The historical developments in the field of plant culture are given in Table 5.2.

Table 5.2 History of plant culture.

Scientist	Year	Contribution/ development
Vochting	1878	first attempted to grow lumps of tissue *in vitro*
Rechinger	1893	observed callus formation by placing beet and candelion roots on wet filter paper
G. Haberlandt	1902	regarded as father of plant tissue culture; first to culture (but not with 100% success) isolated, fully differentiated cells in a nutrient medium containing glucose, peptone and Knop's salts; realized that asepsis is essential for culturing; laid down the requirements for cell divisions
Hanning	1904	succeeded in culturing embryos of *Raphanus* and *Cochlearia*
Lampercht	1918	made attempts to find suitable media and optimal
Kundson	1919	conditions for growth of explant *in vitro*
Nemec	1924	
Knop & Prdifer	-	developed synthetic media
Milliard	1921	developed root culture technique
W. J. Robbins W. Kote	1922	developed a technique for root culture using maize roots
Laibach	1925	first reared hybrid embryos (*Linum perenne* x *L. austriacum*)
Phillip R. White	1937	successfully maintained tomato root culture for 30 years
R. J. Gautheret & P. Nobecourt	1937	successfully cultivated carrot tissues; both P. R. White and R. J. Gautheret are regarded as founders of methods for cultivating excised plant tissues- roots, tips, meristems, cambium
Van Overbeek	1941	used coconut milk for culturing embryos of Dhatura
H.E. Street	1950	established the role of vitamins in plant growth and shoot root relationship
W. H. Muir	1953	devised a technique of agitating callus tissue in liquid medium to obtain single cell culture; and developed 'nurse' culture method
Welmore & Sorokin	1955	established the role of auxin and vitamins as growth controlling factors
F. Skoog &	1955	discovered cytokinins; and
C. O. Miller	1957	obtained roots and stems from callus treated with auxin and kinetin

Table 5.2 *Contd…*

Scientist	Year	Contribution/ development
W. Tabeche	1959	developed submerged suspension culture
G. Morrel	1960	used gibberellins for induction of proliferation and differentiation of meristems- shoot apex culture technique
F.C. Steward	1960	demonstrated the possibility of production of complete plant from a single cell through the development of embryoids from carrot cells in suspension
E. C. Cocking	1960	isolation and culture of protoplasts
	1970	achieved fusion of protoplasts
S. G. Guha & S. C. Maheshwari	1966	produced haploid embryos from cultured pollens and anthers
Cortson	1972	first to produce a somatic hybrid
J. P. Nitsch	1974	cultured microspheres
Bajaj	1976	regenerated plants from cryopreserved plant tissue
Murashige	1977	proposed the production of artificial seeds
G. Melchers	1978	hybridized potato and tomato
K. A. Barton & W. J. Brill	1983	used plasmid vectors for inserting foreign genes into protoplasts
Lazar	1983	first to use protoplast fusion technique to produce cybrids
M. D. Chilton	1983	produced transformed tobacco plants

Further the developments made are related to the manipulation of plants at the genetic level and involve the production of transgenic plants.

5.4   Totipotency: The principle underlying plant culture

The fundamental principle of plant tissue culture is to free the plant part (explant) from its inter-organ, -tissue, or –cellular interaction present *in vivo* and allow it to grow (to any extent: callus, embryo, organ or plantlet) under *in vitro* controlled conditions.

The technique is based on the concept of totipotency i.e. an isolated single plant cell/ tissue (or for that matter any differentiated tissue in general) has the capacity and is able to develop into a whole plant (organism). This involves first dedifferentiation i.e. loss of specialization by the already differentiated somatic cell or tissue to form an amorphous mass of cells called the callus. In the second step, callus is induced to redifferentiate by providing appropriate culture conditions to develop embryoids, plantlets and later onwards the whole plant. The concept is depicted in Figure 5. 1.

This property of totipotency of plant cells/ tissues is in contrast to the animal cells / tissues where only the fertilized egg and pre-implantation 8-cell staged embryonic cells of blastula alone are 'totipotent' and not the other somatic cells.

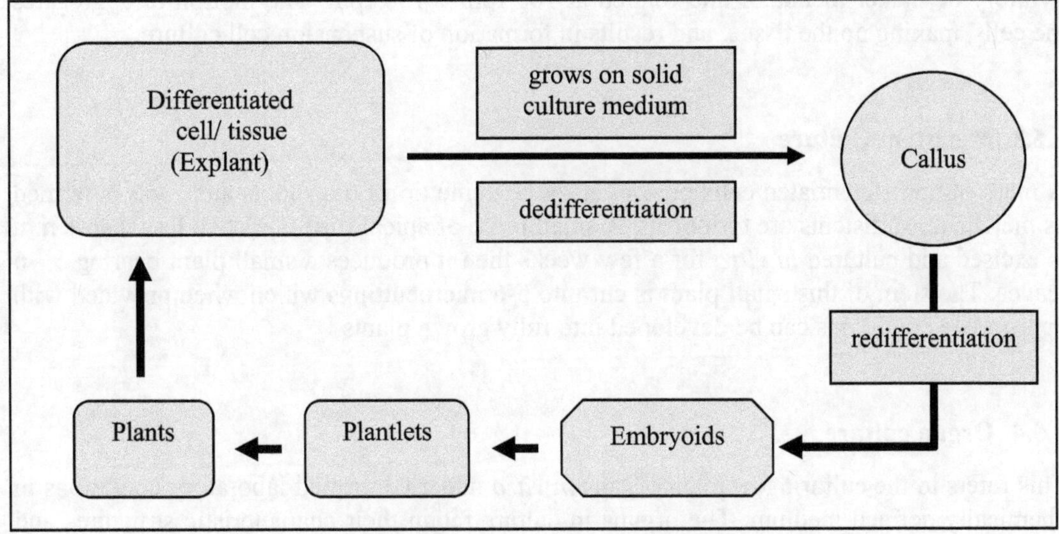

Figure. 5.1 Concept of totipotency.

5.5   Types of Plant Culture

5.5.1 Callus Culture

This is the most common type of culture. The callus (pleural calli) may be defined as an amorphous and loose mass of parenchymatous cells being composed of undifferentiated, unorganized and highly vacuolated cells which develop due to proliferation of cells from parent tissue or an isolated cell/ tissue i.e. the explant, being cultured on solidified medium (called the induction medium). The callus appears like as if a piece of white paper has been ruffled and chewed up. It may be hard (i.e. lignified) brittle or soft to touch. The callus can grow indefinitely without forming any recognizable plant tissues. Plantlets (with roots and shoots) can be generated from the callus by encouraging embryogenesis by transferring the callus (or a piece of it) to a medium having definite levels of plant growth regulators like auxins and cytokinins (maturation medium). Difference between embryogenesis and organogenesis is only that if a piece of tissue is used as the parent tissue to generate the plantlet then the term embryogenesis is used and if a single plant cell is used as the parent cell then the term organogenesis is applied to the technique of generating plantlet.

Generally, in order to induce the formation of shoots, a high ratio of cytokinin to auxin is employed in the medium, while for inducing the formation of roots a high ratio of auxin to cytokinin is employed.

5.5.2 Suspension Cell Culture

This is developed by transferring small cut-pieces from the callus culture into a series of 250 ml Erlenmeyer flasks each containing 50ml of the medium. The flasks are then placed in gyratory or shaker incubator and rotated at 100 rpm – 150 rpm. The motion disaggregates the cells, making up the tissue, and results in formation of suspension cell culture.

5.5.3 Meristem Culture

A mass of undifferentiated cells present at the extreme tip of the shoots and roots is termed as meristem. Meristems are totipotent. A small piece of apical (tip) meristem less than 1 mm is excised and cultured *in vitro* for a few weeks then it produces a small plant bearing 5 - 6 leaves. The stem of this small plant is cut into 5-6 microcuttings which when provided with appropriate conditions can be developed into fully grown plants.

5.5.4 Organ culture

This refers to the culturing of plant organs *in vitro* under controlled laboratory conditions in chemically defined medium. The organs in culture retain their characteristic structures and features unlike callus culture where the tissue looses its organization (i.e. it dedifferentiates).

Organ culture may be classified into two types depending upon the organs:

1. *Vegetative organ culture:* E.g. culture of roots, leaf shoot tip etc.

2. *Reproductive organ culture:* E.g. flower, ovary, ovule, embryo, pollen, seed or fruit culture

5.5.5 Microspore culture

In order to produce a large number of homozygous plants (clone), microspore culture is one popular technique.

Plants, like animals, are heterozygous for alleles i.e. they have two copies of gene in each cell, received from each of the parents. To have a homozygous plant, anthers from male plants are cultured and the haploid cells i.e. cells containing only one set of chromosomes in the anther called microspores, are allowed to form clones from which the entire haploid plant can be grown. However, it is not feasible to grow haploid animals!

The haploid plants growing in culture have only one set of chromosomes which can be doubled to make the normal diploid plant by treatment with colchicines. It is to be noted that both the copies of their chromosomes will be identical i.e. the plant would be homozygous.

5.5.6 Protoplast Culture

Plant, fungal and most bacterial cells possess a very strong, tough, thick cell wall. A cell from which the cell wall has been removed, leaving the cell surrounded only by its limiting plasma membrane is called a protoplast. The reason for preparing plant protoplast is the ease with which a foreign DNA can be introduced into it; otherwise getting the DNA into a cell through the impervious cell wall is extremely difficult. Not only DNA, but large molecules like plasmids, chloroplasts, mitochondria, chromosomes, nuclei, bacteria and other protoplasts can also get into the protoplast relatively easily. All such introductions of foreign elements are necessary to bring about genetic recombination.

Protoplasts are generally prepared by dissolving the cell wall by treatment with enzymes like cellulase and pectinase. The fusion of protoplasts may be spontaneous if the protoplasts are derived from the same species or else it has to be induced if the protoplasts come from two different species. The fusion may be induced by one of the following ways:

1. Treatment with 5.5% w/v sodium nitrate in 10% w/v sucrose solution followed by incubation at 35°C,

2. Treatment with 0.05M calcium chloride in 0.4M mannitol at a pH of 10.5 followed by incubation at 37°C for 35-40 minutes,

3. Treatment with equal volumes of PEG and

4. Application of an electric shock.

The protoplasts have the property to spontaneously regenerate the cell wall in about 24 hours and are totipotent to generate plantlets having improved or new characteristics as encoded by the foreign genetic element.

5.6   The Technique

The culturing of plant cells/ tissues, as mentioned earlier, needs to be carried out strictly under aseptic conditions using sterilized medium, equipments, glassware, explant etc. for want of control of microbial contamination. Further, there has to be provision for controlling the amount of light, temperature and humidity in the culture room during the various phases of growth of the culture. Therefore, proper planning and organization is demanded to set up a plant tissue culture laboratory. Consideration has to be given to the following points:

5.6.1 Laboratory Design

The layout of a typical plant tissue culture laboratory should have the following five distinct areas equipped with necessary equipments:

1. *Media preparation room/ area*

2. *Aseptic transfer room/ area*

3. *Environmentally controlled culture room:* It should have racks with light arrangements, timers, air cooler/ air conditioner for maintaining temperature at 25°C±2°C, incubators having light/ dark photoperiod control, rotary shakers providing speed of 80-220rpm, having capacity of 100ml-250ml, light intensity 2000 to 4000 lux, lux meter etc.

4. *Analytical room:* This is where the properties/ characteristics of growing culture can be determined. It should be equipped with microscopes with photographic attachment, colorimeter, laboratory centrifuge etc.

5. *Acclimatization room:* This is where the plantlets growing *in vitro* are transferred and planted for the first time in soil in pots and kept under controlled conditions like high illumination of 4000-10,000 lux, high humidity 90%-100% etc. This step in plant tissue culture is termed 'soil hardening'

5.6.2 The culture medium: Components

The entire success of plant tissue culture depends upon the appropriate selection and development of a culture medium. In general, the culture medium consists of the following ingredients:

1. *Inorganic salts*

    (i)   Macroelements required in large concentration of mmol/l e.g. nitrogen, phosphorous, potassium, calcium, magnesium and sulphur.

    (ii) Microelements required in small concentration of μmol/l e.g. boron, manganese, zinc, copper, molybdenum, chloride, aluminum.

2. *Organic carbon source*

    (i)   Sugars: E.g. sucrose, glucose required in a concentration of 2%-5% w/w; but these undergo caramelization on heating, react with amino acids, degrade to form melanoidin that may inhibit cell growth.

    (ii) Hexitols: E.g. myo-inisitol. This is considered to have growth promoting and vitamin-like action. Other examples include mannitol and sorbitol.

3. *Vitamins*

    These are required in μmol/l quantities and they serve as co-enzymes. Examples include riboflavin (Vit. B_1), ascorbic acid (Vit. C), nicotinic acid (Vit. B_3) and pyridoxine (Vit. B_6)

4. *Amino acids and amides*

    These play an important role in morphogenesis. Examples include:

    L-arginine for rooting

    L-tyrosine for shoot initiation

    L-serine for induction of haploid embryos

    L-glutamine and L- asparagine for inducing somatic embryogenesis.

5. *Plant growth regulators or Phytohormones*

    There are four classes of plant growth regulators. Which growth regulator and in what concentration it is to be included in the medium is decided by the type and purpose of culture. The four classes of plant growth regulators are given in Table 5.3.

Table 5.3 Use of plant growth regulators in plant tissue culture.

Plant growth regulator	Use
Auxins	
Indole-3-butyric acid (IBA)	induction of cell division,
Naphthaleneacetic acid (NAA)	root inititation,
p-Chlorophenoxyacetic acid (p-CPA)	callus induction
Naphthoxyacetic acid (NOA)	
Trichlorophenoxyacetic acid (2 4 5-T)	
Dichlorophenoxyacetic acid (2 4-D)	
Indole acetic acid (IAA)	
Cytokinins	
Benzyl amino purine (BAP)	adenine derivatives;
Isopentyl adenine (2 IP)	promote cell division,
Furfurylaminopurine (FAP)	shoot proliferation,
N-methlaminopurine (MAP)	organogesis, somatic embryogenesis
Gibberlins	
GA$_3$	used in meristem culture for plant regeneration and elongation
Abscisic acid	used in embryo culture and somatic embryogenesis

6. *Natural complex extracts*

These are included in the media for their growth promoting properties. Examples include coconut milk, yeast extract, malt extract, tomato juice, potato extract, casein hydrolysate, fish emulsion etc.

7. *Gelling agents*

These are used for preparing solid media. Example includes agar.

8. *Antibiotics*

These are included for control of bacterial and fungal growth and are generally included only when the explant is heavily contaminated. Examples include penicillin, streptomycin etc.

9. *Antioxidants and adsorbents*

These are employed to prevent darkening (browning) of explant in culture. Examples include citric acid, ascorbic acid, pyrogallol, phloroglucinol, L-cysteine, PVP and activated charcoal.

10. *Water*

The water employed for preparing the medium should be demineralized and double distilled.

For a typical plant culture medium the pH of 5.6 to 5.8 is adjusted using 0.1M NaOH/HCl.

The basic culture medium i.e. one which contains mineral salts is sterilized by autoclaving. Rest other ingredients which are thermolabile such as carbohydrates, vitamins, growth hormones and plant extracts are sterilized by filtration and added aseptically to the sterile basic medium.

There are numerous examples of basic plant tissue culture media, but Murashige-Skoog (MS) medium is the one that is almost always used.

5.6.3 The explant: Surface sterilization

Explant is a piece of plant body that has been detached and is to be transferred to a culture medium for culturing *in vitro*. The explant, collected from the plant from its natural surroundings, will have its surface contaminated (with microbes) and sometimes with the possibility of heavy contamination. As mentioned earlier, the entire culture technique needs to be carried out aseptically using sterile medium, glassware etc., therefore, the prime step of foremost importance in getting started with the culture is to sterilize the surface of the material. This is known as surface sterilization.

In order to carry out surface sterilization, the explant is submerged in a solution of a chemical agent. It is important to note that the selection of the type of chemical agent, its concentration and time of exposure/ submergence of explant are very crucial factors. Since, our aim is to kill the microbes on the surface of the explant, at the same time if a high concentration or a long exposure time is used then there is all likely possibility that the explant looses its viability. On the contrary, if too low a concentration or too short exposure time is employed then there is a possibility of microbes surviving and subsequent contamination. In Table 5.4, the commonly employed surface sterilizing agents are listed.

Table 5.4 Commonly employed surface sterilizing agents.

Chemical agent	Concentration (%w/v)	Exposure time (min.)
Benzalkonium chloride	0.01-0.1	5-20
Calcium hypochlorite	9-10	5-30
Ethyl alcohol	70-95	1-5
Mercuric chloride	0.1-1.0	2-10
Hydrogen peroxide	3-12	5-15
Silver nitrate	1	5-30
Sodium hypochlorite	0.5-5	5-30

After exposure to the sterilizing agent, the explant must be given 2-3 rinses with sterile water so as to completely wash out the sterilizing agent, which may otherwise be inhibitory to the growth of the explant.

The plant culture technique is shown schematically in a nut shell in Figure 5.2.

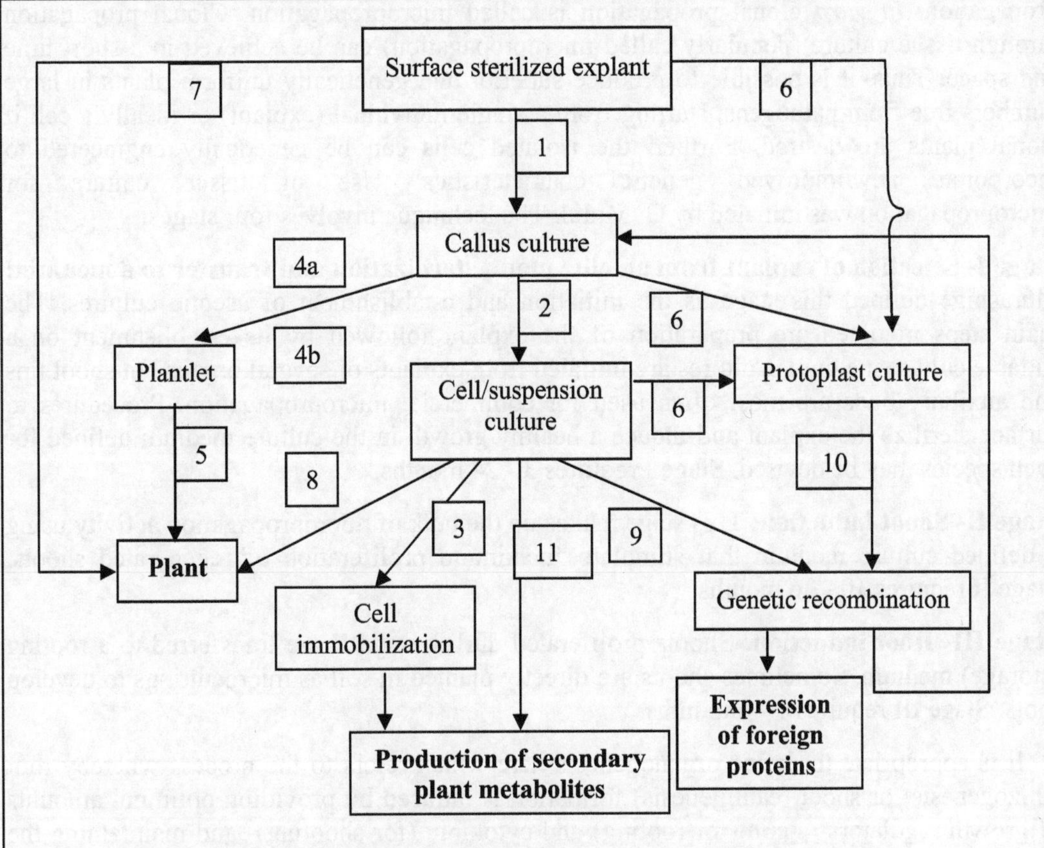

1: Solid medium having low concentration of auxin and cytokinin
2: Less than 1mm pieces of callus in liquid medium agitated at 125 rpm
3: Immobilization using calcium alginate
4a: Solid medium having low auxin and high cytokinin concentration for shoot induction
4b: Solid medium having high auxin and low cytokinin concentration for root induction
5: Soil hardening
6: Enzymatic or mechanical methods for preparing protoplasts
7: Micropropagation, where explant is shoot tip, auxiliary bud
8: Clonal propagation where explant is single cell
9: Genetic engineering methods (see Section 5.7.2)
10: Introduction of foreign genetic element
11: Methods for enhancement of secondary metabolite production (see Section 5.7.3)

Figure. 5.2 The plant culture technique.

5.7   Applications

The applications of plant tissue culture are presented under the following four headings:

5.7.1 Cloning or micropropagation

The multiplication of genetically identical copies by asexual means is called clonal propagation. *In vitro* clonal propagation is called micropropagation. Clonal propagation through tissue culture (popularly called micropropagation) can be achieved in a short time and space. Thus, it is possible to produce superior and genetically uniform plants in large numbers free from pathogens, starting from a single individual (explant) or ideally a cell if clonal plants are desired. Further, the isolated cells can be genetically engineered to incorporate new/improved genetic characteristics. Use of tissue culture for micropropagation was initiated by G. Moral. The technique involves four stages:

Stage I- Selection of explant from an elite plant, sterilization and transfer to a medium: Murashige defined this stage as the initiation and establishment of aseptic cultures. The main steps involved are preparation of the explant followed by its establishment on a suitable culture medium. Cultures are initiated from explants of several organs but shoot tips and auxiliary buds are most often used for commercial micropropagation. Procedures to surface sterilize the explant and induce a healthy growth in the culture medium defined for each species may be devised. Stage I requires 3 - 24 months.

Stage II- Shoot induction: This stage makes up the bulk of micropropagation activity using a defined culture medium that stimulates maximum proliferation of regenerated shoots. Stage II requires 10 - 36 months.

Stage III- Root induction: Shoots proliferated during stage II are transferred to a rooting (storage) medium. Sometimes shoots are directly planted in soil as microcuttings to develop roots. Stage III requires 1 - 6 months.

It is appropriate to define organogenesis here which refers to the process whereby root (rhizogenesis) or shoot (cauligenesis) formation is induced by providing optimum amounts of growth regulators- auxin (for rooting) and cytokinin (for shooting)- and maintaining the required conditions of light- red for rooting and blue for shooting- and a temperature of 33°C.

Stage VI- Transplantation of plants to soil: Steps taken to ensure successful transfer of the plantlets of stage III from the aseptic environment of the laboratory to the environment of the green house comprise stage IV. Unrooted stage II shoots are also acclimatized in suitable compost mixture or soil in the pots under conditions of light, temperature and humidity inside the greenhouse. In stage IV, it takes 4-16 weeks for the finished product to be ready for sale or shipment.

The technique has been employed for mass production of nursery stock species, ornamental plants (floriculture), vegetable and fruiting plants (banana, grapes) and field crops. Since the technique is skill intensive and expensive than conventional vegetative breeding, it is basically used for propagating valuable and slow-growing plants.

5.7.2 Protein production in plants: Genetically modified (GM) plants

These days plants are fast becoming the most favoured, fascinating and attractive organisms for the bulk production of foreign (animal) proteins. This attraction for using plants as hosts in comparison to bacterial or animal cells is attributed to the following reasons:

1. Plant media do not include special ingredients like serum which is a must for culturing animal cells,
2. Large amounts of storage proteins are accumulated and, therefore, can be produced in seeds or tubers,
3. Plants are eukaryotic like animals, therefore, expression of genes (exons with introns and post-translational changes) are quite similar which is not achieved in bacterial hosts,
4. Plant production i.e. agriculture as industry, is the largest industry in the world, hence suitable for production of biotech based drugs and
5. There are no issues about contamination of the product with virus, bovine serum albumin or bacterial toxins.

However, in order to genetically engineer the plant so as to form transgenic plant, the desired foreign DNA has to be introduced into the plant cell. The biggest hurdle of getting the DNA into a plant cell is the presence of the robust cell wall. Out of the numerous methods available for this, one of the methods that is used involves the use of Ti plasmid as the vector and is discussed below.

5.7.2.1 Ti plasmid of *Agrobacterium tumefaciens*: *Agrobacterium tumefaciens* is a Gram-negative, rod shaped, aerobic, soil bacterium that infects dicotyledonous plants by getting adsorbed on the surface of plant cells exposed at wound sites. It is responsible for causing a neoplastic disease called 'crown gall'. Crown refers to the junction between the root and shoot i.e. the point where the plant is most likely to get wounded. It is at this site (crown) where the galls develop i.e. a hard, lumpy, tumourous growth on the plant.

There are two distinguishing characteristics of the cells infected by this pathogen:

• The transformed cells can grow *in vitro* in a culture medium that is devoid of the much required phytohormones – auxin and cytokinin and

• The transformed cells synthesize amino acid polymorphs or sugar derivatives referred to as 'opines'. There are two kinds of opines- octopine and nopaline – produced as a result of infection by the two different strains of *Agrobacterium tumefaciens*.

Based on the experiments conducted during the 1970 s by various research groups, it became evident that *Agrobacterium tumefaciens* contains a large extrachromosomal element (more than 200kb) designated as Ti plasmid (Ti for tumour inducing) which is responsible for cell transformation and production of the disease.

Briefly, the Ti plasmid consists of the following segments (see Figure 5.3 as you read along):

1. **T-DNA (for transfer DNA):** This is the portion of Ti plasmid that gets excised from the plasmid and is actually transferred to and gets integrated with the plant genome. It is made of 23 kbp and carries some thirteen genes encoding for the synthesis of opine

(either nopaline or octopine) and also for the enzymes (tryptophan mono-oxygenase ($T_{ms}1$), aminohydrolase ($T_{ms}2$) and dimethyl allyl transferase T_{mr}) that catalyze the synthesis of phytohormones responsible for the uncontrolled, tumour formation.

2. **Light and left border sequences**: The T-DNA region is bordered on both the sides (left and right) by sequences consisting of 25bp, which are responsible for the transfectivity of Ti plasmid.

3. **Vir region (for virulence):** This contains genes that encode information for DNA processing enzymes required for the excision, transfer and integration of T-DNA segment.

   Normally, these genes have a low level of expression but when the bacterium comes in contact with the wounded plant cells or plant exudates rich in phenolic compounds like acetosyringone, these genes are induced and expressed at high levels thus leading to infection.

4. **Opine catabolism region**: The opines produced by the bacterium are used itself as source of energy. Thus this region encodes for enzymes that can catabolize opines.

5. **The origin of replication region**: This is the unique site from where the replication starts.

This natural mechanism of transfection involving the transfer and integration of T-DNA from Ti plasmid of *Agrobacterium tumefaciens* can be employed as a means/ vector of transferring the foreign gene attached with T-DNA into a plant cell having a rigid cell wall, making the transference of a foreign gene easy which otherwise is a difficult task. For this, the Ti plasmid is engineered such that the foreign gene is transferred to the T-DNA region (adjacent to the nopaline synthesis genes) which shall carry the foreign gene along with as it infects the plant. It is important to note that this engineering is done by what is referred to as homologous recombination rather than by the conventional rDNA technique. It is also to be noted that the transfer of foreign gene using *Agrobacterium tumefaciens* is possible only in the dicots and not in monocots, since the bacterium infects only the dicots.

5.7.2.2 Other methods: Other methods of gene transfer in plants include:

1. Electroporation
2. Microinjection
3. Gene gun (biolistics)
4. Phage mediated (e.g.cauliflower mosaic virus) and
5. Protoplast transformation

These methods have been discussed in section 2.7.2.4 of this book.

5.7.3 Production of secondary plant metabolites

Secondary plant metabolites are the compounds that are produced during the stationary phase of growth and are not required by the plant for its growth and development. Among the secondary plant metabolites, there are compounds that have pharmacological activity and so have an immense potential in human health. Apart from being used as drugs, secondary plant metabolites are also used as flavorants and fragrants.

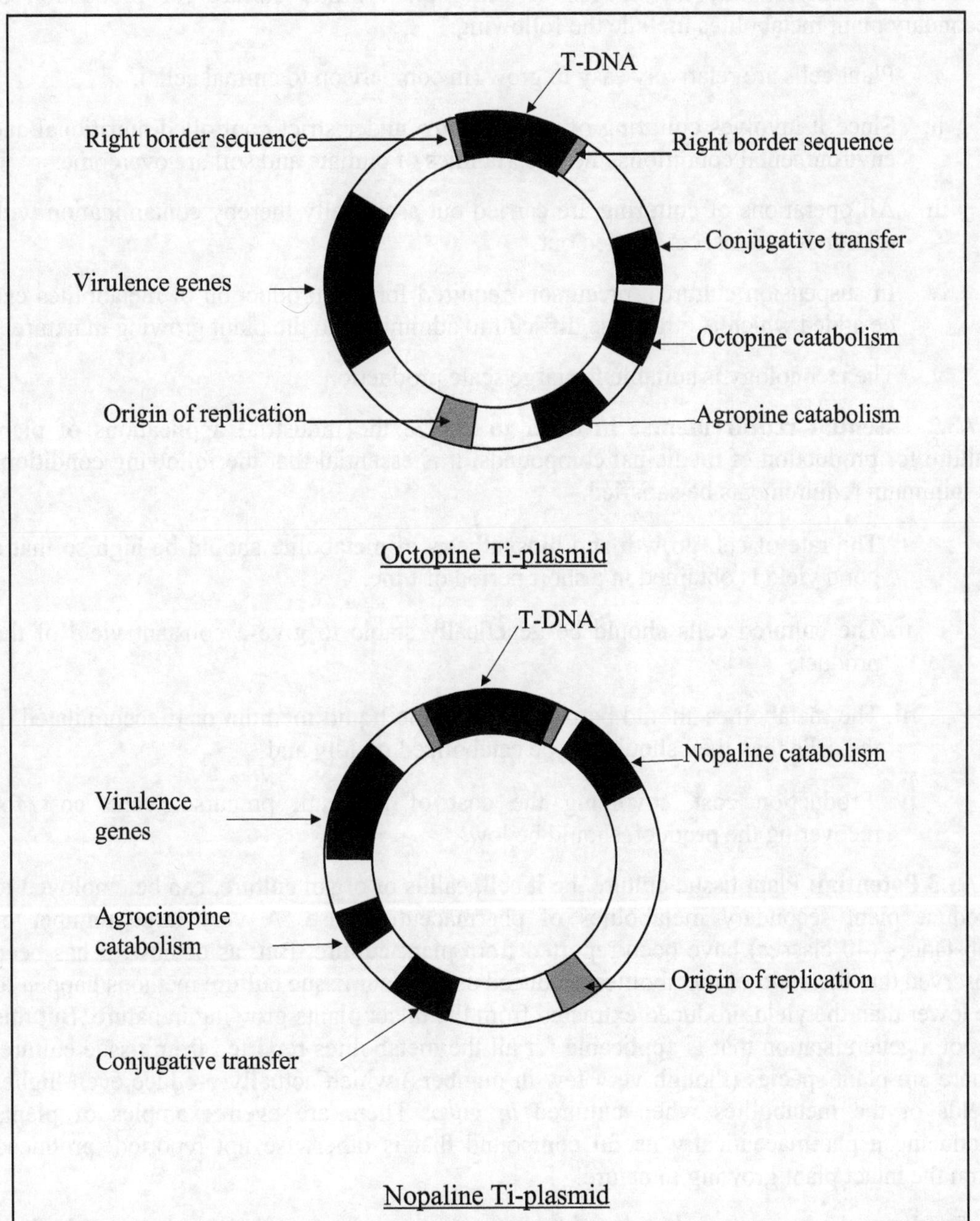

Figure. 5.3 Diagrammatic presentation of Ti plasmids.

5.7.3.1 Advantages: The advantages of using plant tissue culture for production of secondary plant metabolites include the following:

i.   Plant cells are relatively easy to grow (in comparison to animal cells).

ii.   Since it involves culturing of cells *in vitro* under strict controlled nutritional and environmental conditions, the uncertainties of climate and soil are overcome.

iii. All operations of culturing are carried out aseptically thereby contamination with microbes and insects is ruled out.

iv.   In suspension cultures, precursors required for the production of metabolites can be added which is otherwise difficult to administer to the plant growing in nature.

v.   The technology is suitable for large scale production.

5.7.3.2 Essential requirements: In order to realize the industrial applications of plant culture for production of medicinal compounds, it is essential that the following conditions as minimum requirements be satisfied:

i.   The rate of cell growth and biosynthesis of metabolite should be high so that a good yield is obtained in a short period of time,

ii. The cultured cells should be genetically stable to give a constant yield of the product,

iii. The metabolites should be released into the liquid medium or if accumulated in the cells then they should not be catabolized rapidly and

iv. Production cost, involving the cost of medium, precursors and cost for recovering the product, should be low.

5.7.3.3 Potential: Plant tissue culture, be it cell, callus or organ culture, can be employed to produce plant secondary metabolites of pharmaceutical use. A very large number of substances (40 classes) have been reported from plant culture. But, as of now, it has been observed that the yield of metabolites produced using plant tissue culture methods happen to be lower than the yield produced/extracted from the intact plants growing in nature. But this is not a generalization that is applicable for all the metabolites reported from tissue culture. There are plant species (though very few in numbers) which actually produce even higher yields of the metabolites when cultured *in vitro*. There are even examples of plants producing a pharmaceutically useful compound that is otherwise not reported/ produced from the intact plant growing in nature!

So, the explant in culture does have the potential to produce secondary plant metabolites, what is needed is to further and better understand the biosynthetic pathways and find ways and means to enhance the production of desired metabolites.

5.7.3.4: Methods for enhancing the production of secondary plant metabolites: Ways have been applied as to how the production of secondary plant metabolites can be enhanced. These include:

1. **Addition of precursors in culture medium:** For example addition of phenylalanine and ornithine result in increased production levels of alkaloids produced from cultures of *Solanaceous sps., Catharanthus roseus, Cinchona ledgeriana and Capsicum frutescence*. Likewise, inclusion of cinnamic acid and tropic acid enhance the production of flavanoids and tropane alkaloids respectively in cultures.

    However, it is to be noted that an optimum concentration of the precursors has to be used; a high concentration can be lethal to the cultures.

2. **Alteration of culture conditions:** Unlike primary metabolites the production of secondary metabolites depends very much on the environmental and culture conditions which dramatically alter the yield of the metabolites. Light is one such factor. For example light is found to have a stimulatory effect on the production of secondary metabolites such as volatile oil, flavone, flavanoids flycosides, polyphenols, plastiquinone, flavanoids and cartenoids. Light may also have inhibitory effect such as in the production of shikonin.

3. **Inclusion of optimum concentration of nutrients:** Appropriate levels and type of nutrients, growth factors, vitamins etc. when included in culture medium increase cell growth and production of metabolites. For example, production of shikonin is enhanced if medium contains a very high concentration of sucrose, while a very low concentration of sugar in medium increases the production of ubiquinone-10. Likewise, combination of cytokinin and auxin in appropriate concentrations has a synergistic effect on the production of steroids in cultures of *Trigonellum foenum-graceium*.

    Take another example, in tobacco cultures, 2,4-D inhibits the synthesis of nicotine while inclusion of kinetin enhances its production. Likewise, inclusion of 2,4-D and NAA has an inhibitory effect on the synthesis of shikonin in cultures of *Lithospermum sps*. Further, NAA promotes the synthesis of anthraquinones in *Morinda* cultures. 2,4-D has a stimulatory effect on the synthesis of L-DOPA while it does not effect the synthesis of anthraquinones in cultures of *Cassia tora*.

4. **Somaclonal variation and artificial mutation:** The cells growing in cultures undergo extensive genetic changes. For industrial production, a variant (by nature) cell (or its clone) can be selected and used that produces high yields of the desired metabolite. Alternatively, the cultured cells can be artificially (by man) mutated by employing chemical mutagens and the mutants producing high yields can be selected for production of secondary metabolites.

5. **Morphogenesis:** Certain compounds are synthesized and stored in specific parts of plant. For example, essential oils are stored in sex glands/ ducts, tropane alkaloids in roots, latex in lactiferous glands etc. For such compounds, it is necessary to induce organogenesis so that the compound can be synthesized and stored in that organ just as in the natural intact plant.

    Root induction in cultures of *Scoplia parviflora* and induction of organogenesis in *Datura innoxia* is found to result in increased production of alkaloids.

However, shikonin, (one of the very few products, the production of which by plant culture is a true commercial story) is found in cork cells in the intact plant. But, its suspension culture (that means without organogesis) is found to produce very high yields of shikonin!

6. **Use of elicitors**: This method and the next one using Ri plasmids, are the two methods that are well established, accepted and have a profound influence for enhancing the production of secondary plant metabolites.

Elicitors are specific compounds or mixtures which when added to the suspension cultures stimulate the production of metabolites. Commonly these are known as 'stress agents' i.e. the plant produces more compound under stress when it is challenged by an external agency. Elicitors are classified into two types:

1. *Abiotic elicitors*

- Physical agents e.g. u.v radiation, pH, osmotic pressure, heavy metal ions etc.

- Chemical agents e.g.sachharides, glycoproteins, arachidonic acid etc.

2. *Biotic elicitors:* these are complex homogenates or filtrates and extracts of:

- Bacterial origin e.g. of *Botrytis* sp, *Phytophthora* sps etc.

- Fungal origin e.g. of *Aspergillus,* micromucor, yeast etc.

The use of elicitors for product accumulation is given in Table 5.5.

Table 5.5 Representative examples of elicitation of product accumulation.

Biotic elicitor	Cell culture	Product
Aspergillus niger	*Cinchona ledgeriana*	Anthraquinones
	Rubia tinctoria	
	Morinda citrifolia	
Botrytis sp	*Papaver somniferous*	Sanguinarine
Chrysosporium palmorum	*Catharanthus roseus*	Tryptamine, Ajmalacine, Catharanthine
Micromucor isabellina	*Catharanthus roseus*	Catharanthine
Pythium aphanidermatum	*Catharanthus roseus*	Strictosidine, Ajmalacine, Taberoonine, Lachnuricine
Yeast	*Glycine max*	Glyceolline

7. **Hairy root culture**: Ri plasmids of *Agrobacterium rhizogenes* – *Agrobacterium rhizogenes* is a pathogenic soil bacterium that infects dicotyledonous plants and induces the proliferation of roots at the site of infection (compare with Ti plasmid of *Agrobacterium tumefaciens*). The plant is actually transformed by the bacterium i.e. the bacterium inserts T-DNA (T for transfer) carried on its Ri plasmid (Ri for root inducing) into the plant genome. The transferred portion (T-DNA) of Ri plasmid carries genes which code for auxin synthesis and other rhizogenic functions. Such a transformation can be established *in vitro*, wherein the explant (a detached portion of plant, scratched midrib of leaf or stem of plantlet) in culture is inoculated with a suspension of *Agrobacterium rhizogenes*. Now either a callus may develop first and then there is profuse induction of roots (in some species) or straight away the roots may appear (in most of the species) at the site of infection. It takes about 1-4 weeks to establish such a hairy root culture. Inclusion of acetylsyrigone, a phenolic compound, helps bring about increased infectivity.

The hairy root cultures are potentially applicable to the production of root derived secondary plant metabolites from dicot plants. Examples include the production of tropane alkaloids, atropine, hyosyamine, solasidine, catharanthus alkaloids etc.

Examples wherein hairy root cultures have enhanced the yield of metabolites include the production of betaine from *Beta vulgaris* and hyoscyamine from *Datura stramonium*. Also, in comparison to the selection (itself a difficult task) of variants and mutants, hairy root cultures have a prolonged genetic stability.

5.7.4 Miscellaneous

Plant tissue culture apart from being used for the production of secondary plant metabolites is also useful in the following ways:

1. Storage of germplasm: In this case identical clones are generated by tissue culture methods and the plant material/ tissue is stored at low temperature using liquid nitrogen termed as 'cryopreservation'.

2. Production of artificial seeds: This is nothing but the tissue culture generated somatic embryos, surrounded by nutrient medium and encapsulated in a synthetic membrane made of calcium alginate, polyoxyethylene or polyacrylamide gel.

3. Production of triploid plants used to produce seedless fruits

4. Physiological and plant disease studies

<table><tr><td>6</td><td></td></tr></table>

6   ANIMAL CULTURE

CHAPTER CONTENTS AT A GLANCE

6.1   Introduction

Animal culture includes cultivation of animal cells, tissues or organs. Animal cell culture relates to the cultivation of cells from a multicellular organism usually mammalian cell or even more commonly insect cell, in a suitable nutrient medium aseptically under carefully controlled conditions. Animal culture includes:

1. Cell cultures-There are of four types of animal cell cultures:

   i.   *Primary cell culture:* This culture is derived from normal tissue taken from an animal such as mouse, hamster, chicken, monkey or human. This tissue is treated with proteolytic enzymes to give a cell suspension. These cells are then placed along with the growth medium in a vessel and incubated. The cells are observed to grow as a monolayer on the surface of the vessel. This is referred to as primary culture.

   ii.   *Secondary cell culture or cell line:*   The primary culture or the original tissue contains many (heterogeneous) types of cells and also have variable growth functions. Sub culturing a fraction of cells from the primary culture yields a more homogenous culture of cells. This culture is called a cell line, which can be propagated, characterized and stored.

iii. ***Diploid cell strain:*** In the cell line culture, still there may be several 'cell lineages'. Among the various cell lineages, a particular cell lineage with specific properties is identified from the bulk of cells and then cultured. This is then described as cell strain. In other words, cell strain is derived from a single cell type with specific stable properties like having marker chromosomes, marker enzymes or antigens.

iv. ***Continuous cell line:*** All the above described cultures, if derived from a normal tissue can be sub-cultured only for a limited number of times with the cells generally dividing for 20-100 times (i.e. Hay Flick limit) and the cells in culture finally die out. This is called cellular senescence and may be connected to aging process. Thus, the culture from a normal tissue is finite.

But if cells are obtained from a malignant tissue or if they are mutated, then such cell lines are capable of undergoing infinite number of divisions in culture i.e. they are "immortal". This is referred to as continuous cell line.

The classical example of this is the HeLa cell line named after Henrietta Lacks, a 31 year old black woman suffering from cervical cancer. In 1951, this cancerous tissue was cultivated in the laboratory and exposed to gamma radiations but the tumorous cells in culture continued to grow. These cells are now named HeLa cells.

Apart from cell culture, animal culture also includes:

2. **Tissue Culture :** wherein fragments of excised tissue are grown in culture medium and

3. **Organ Culture :** wherein the entire embryos, organs or tissues are excised from the body either by vivisection or shortly after brain death and cultivated *in vitro*.

6.2   Properties of the Animal Cell

The animal cells are of two types:

1. **Anchorage dependent** i.e. cells which remain viable and will proliferate when attached to a solid substrate eg: primary cultures, normal cell strains etc. or

2. **Anchorage independent** i.e. cells which will proliferate in suspension eg: transformed cells, tumor cells, hybridomas. Table 6.1 presents a brief compilation of the properties of animal cells.

Table 6.1 Properties of animal cells.

Size	:	10-100 microns
Shape	:	Spherical in suspension
Cell wall	:	Absent
Plasma membrane	:	Thin, fragile and shear sensitive and
		composed of phospholipids and proteins.

Phospholipid molecules are arranged in two parallel rows called bilayer as:
- Polar heads: phosphate group+ glycerol (hydrophilic)
- Non-polar tails: fatty acid (hydrophobic)

Proteins are of two types-

• Peripheral	:	Involved in chemical reactions and can be easily removed
• Integral	:	Channels for exchange of substances and not easily removed
Nucleus & organelles	:	Membrane bound

Many organelles	:	Endoplasmic reticulum, golgi apparatus, mitochondria, lysosome
Surface	:	Negatively charged
Grow on	:	Positively charged surfaces, e.g. collagen

6.3   The Technique of Animal Cell Culture

Advancements in two fields of medical research, namely production of viral vaccines and understanding of neoplasia, helped develop the modern cell culture technology. Other factors like development of reliable sera and media, use of antibiotics and use of clean air equipment and pressure from groups fighting for animal right further hastened the progress especially in cell culture.

The animal cell culture technique is developed keeping the above cell properties in mind. The technique is discussed below.

6.3.1 The substrate/ surface: As mentioned most of the vertebrate cells need to spread out on a substrate/ surface on to which they proliferate. The substrates / surfaces used include:

1. Glass e.g. slides, cover slips, flask or test tube.
2. Plastic e.g. sterilized disposable plasticware made of polystyrene, PVC, PTFE, poly carbonate etc.
3. Palladium and stainless steel discs

To improve cell attachment, the surfaces are generally treated with collagen, polylysine, fibronectin (a globulin) or application of feeder layer i.e. a monolayer of special kinds of cells (such as mouse embryo fibroblasts, normal fetal intestine, glial cells etc).

6.3.2   The Culture Medium:

There are two classes of media used for culturing animal cells:

6.3.2.1   Natural substances: These include:

1. *Blood plasma* e.g. plasma from adult chicken. Plasma is used in culture medium to provide:

   - Nutritive substrate
   - Supporting structure
   - Means of conditioning the glass surface for better attachment
   - Protection from traumatic damage of cells during sub-culturing
   - Protection from sudden changes in environment when fluid is changed
   - Localized pocket of medium around cells

2. *Blood serum* e.g. human placental cord serum, chicken serum, mammalian serum (such as fetal calf serum)

Though chemically defined media have been developed, but it is found that they do not support the growth of cells unless they are supplemented with 5%-20% of serum. Thus, serum provides growth factors and physical conditions for growth that are deficient in synthetic media.

Efforts are going on to develop a serum free medium. This requires a complete knowledge of the constituents present in the serum which can be added to the synthetic medium. A partial list of the constituents includes: selenium, transferrin, albumin, globulin, insulin, androgen, hydrocortisone, estrogen, platelet derived growth factor and various other GFs amino acids, glucose, acids and lipids.

3. *Tissue extracts* e.g.: chick embryo tissue extract. The active fractions of the extract responsible for promoting cell growth have been isolated and identified to be the nucleoproteins (especially ribonucleic acids) and the glycoproteins.

4. *Complex natural media*: These are made up of a number of natural occurring media of undefined composition, like beef embryo extract, horse serum, bovine amniotic fluid, yeast extract, betalbumin hydrosylate etc. e.g.: supplemented Hanks-Simms medium, supplemented bovine amniotic fluid medium, serum supplemented yeast extract medium, serum supplemented betalbumin hydrosylate and yeast extract medium.

5. *Biological fluids*: e.g. amniotic fluid, ascetic and pleural fluids, aqueous humor (from eye), insect hemolymph.

6.3.2.2 Chemically defined culture media: Though natural media are mostly used, they have the advantage of not giving reproducible results and their exact composition is unknown. So various scientists, (Correl and Baker 1926; Vogelaur and Erlichman 1933; White 1946; and Morgan et al 1950; eagle 1959) made attempts to derive chemically defined media of known composition.

Almost all the chemically defined media contain:

1. Glucose as galactose
2. Essential amino acids especially glutamine
3. Vitamins of B complex, choline, inositol
4. Inorganic salts- Na^+, K^+, Mg^{2+}, Ca^{2+}, Cr^+, S, Ca^{2+}, HCO_3^{--}
5. Organic supplements- proteins, nucleosides, pyruvates, lipids

Additionally, the chemically defined synthetic media have to be fortified with homologous or heterogeneous serum in a concentration of 5%-20%. The optimum pH of the medium should be 6.9 to 7.8 and should be known and established for each cell line.

Antifungal and antibacterial antibiotics (streptomycin, penicillin) are also added to check the contaminating microorganisms. Some of the chemically defined culture media with their applicability are given in Table 6.2.

Table 6.2 Some chemically defined animal culture media and their applicability.

Cell/tissue/organ culture	Suitable medium
Chick embryo fibroblasts	Eagle's MEM
Chinese hamster ovary (CHO) cells	Eagle's MEM
Chondrocytes	F12
Continuous cell lines	Eagle's MEM
Endothelium	DME, 199, MEM
Glial cells	MEM, SF 12
HeLa cells	Eagle's MEM
Haemopoetic cells	RPMI 1640, Fischer's
Human tumours	SF 12, L 15, RPMI 1640, DME
Dog kidney epithelium	DME
Mouse embryo fibroblasts	Eagle's MEM
Mouse leukemia	Fischer's, RPMI 1640
Mouse myeloma	DME, RPMI 1640
Skeletal muscle	DME, F 12

Abbreviations: MEM-minimum essential medium; F12-Ham's F12; DME- Dulbecco's modified essential medium; SF 12-Ham's 12 plus Eagle's essential and non-essential amino acids; RPMI 1640- Roosevelt Park Memorial Institute;

6.3.3 Explant: Source of Tissue

The excised portion of the animal body that is to be cultivated is called the explant. The source of the explant includes any one of the following:

1. *Embryo-* consisting of cells termed as precursors/ stem/ master cells. The cells are characterized by: low level of specialization, high proliferation capacity and high survival rate.

2. *Adult tissue/ organ-* These may be of three types:
    i)   Stem cell containing-tissues that undergo continuous removal *in vivo* e.g. intestinal epithelium, epidermis, haemopoetic cells, vaginal endometrium etc.
    ii) Committed precursor cell containing-tissues that renew under stress conditions E.g. muscle, gilial, fibroblasts, etc.
    iii) Adult cells – which are characterized as specialized (differentiated) cells and have low growth rate.

3. *Neoplastic-* These are characterized by their uncontrolled growth rate.

The tissues should be collected under aseptic conditions (embryonic tissue) or sterilized (all other tissues) after collection. This is achieved by washing the tissue in balanced salt solution (BSS) containing 100 units of penicillin and 0.5mg/ml of streptomycin.

6.3.4 Cultivation

The sterilized explant tissue as prepared above is further cut into small pieces of 1-2 mm^3. These pieces of tissue may be cultivated as such or are disaggregated to give a suspension of cells which can be cultivated.

6.3.4.1 Primary explant technique

This is a technique for cultivating small pieces of tissue especially skin biopsies. The technique was first developed by Harrison (1907) and Correl (1912). The various explant techniques include:

1. Slide or cover slip cultures (Harrson, 1907).
2. Correl flask cultures (Correl, 1972).
3. Roller test tube cultures.

The explant technique involves embedding the explant in one drop or one ml of blood plasma placed over a cover slip or flask and adding a drop of a mixture of 11 ml embryo extract mixed with serum over it. A glass slide is then inverted over the cover slip and then turned over (as in the hanging drop technique).

6.3.4.2 Disaggregation of tissue

This is achieved by any of the following methods:

1. *Enzymatic disaggregation*:

A tissue may be treated with certain enzymes such as trypsin or pepsin to break up into cells.

Treating the tissue with a solution of crude trypsin i.e. trypsinization, is of two types: cold trypsinization (exposure temperature of 4°C) and warm trypsinization (exposure temperature of 36.5°C). Complete disaggregation of the tissue takes about 3hours-4 hours. Filtration through fine muslin cloth is done every half hourly to obtain the free cells in filtrate from which the cells are recovered by centrifugation. The cells so obtained can now be inoculated in the appropriate medium.

2. *Mechanical disaggregation*:

This involves forcing the tissue repeatedly through:

- A stainless steel sieve (of 1mm and lower aperture size) followed by use of sieves with gradually reducing aperture size or
- A needle in a disposable syringe or
- A pipette.

The suspension of cells so obtained is cultured.

In order to maintain the primary culture as established above, it would be necessary to sub – culture the cells in fresh medium at the appropriate intervals of time. This will also be necessary to obtain a cell line and cell strain. The cells growing in primary culture may be anchorage dependent (source of tissue 1 & 2) or anchorage independent (source of tissue 3). If the cells are adherent to the substrate, they have to be treated with trypsin or other enzyme to dissociate the cells in the monolayer. These cells are then transferred into fresh medium.

For the growth of cells, tissues or organs gaseous oxygen and carbon dioxide are required. The concentrations of both the gases are to be monitored. Carbon dioxide influences the pH and bicarbonate ion concentration of the culture. Maintenance of sterility and aseptic conditions are the key to successful cultivation of animal cells. Since, entry of even a single microorganism (bacteria or fungi), which grow at a faster rate than animal cells, will rapidly grow in the medium and will soon outnumber (though) the large number of animal cells initially inoculated. Apart from utilizing the nutrients, the microbes will produce metabolic wastes and poisonous toxins. So, rigorous sterility is critical in cell culture.

6.3.4.3   *In vitro* vs *in vivo* environment: In comparison to the environment offered *invivo*, the animal cells experience a different environment *in vitro* w.r.t. the following:

- Cell-cell interaction is reduced *in vitro*
- Cell -matrix interactions are reduced *in vitro*
- Nutritional milieu is changed *in vitro*
- Spreading is increased *in vitro*
- Migration is increased *in vitro*
- Proliferation is increased *in vitro*

6.4   Cell Culture for Production

This pertains to the culturing of mammalian cells for producing commercially useful materials. The culture may be setup at small scale in roller bottles of 100ml capacity or at large scale in fermentors or bioreactors of 100's of liters of capacity. A factor that is to be kept in mind while designing the fermentor is that the animal cells are fragile and so gentle methods of agitation (air lift, hollow fiber) are to be used.

6.4.1 Advantages of animal cell culture

The advantages of the use of animal cell culture for production include:

- There is absolute control of physical environment'
- Homogeneity of sample
- Less compound needed than in animal models.

6.4.2 Disadvantages of animal cell culture

The disadvantages of the use of animal cell culture for production include:

- Animal cells are hard to maintain in culture,
- Only a small amount of tissue can be grown that too at high cost,
- Dedifferentiation of animal tissue,
- Instability and aneuploidy in animal cells,
- Low productivity of desired product
  - Good producers: produce 100 µg/ml/day
  - Average producers : produce 20-100 µg/ml/day
  - Poor producers : produce < 20 µg/ml/day

- Animal cells have a slow growth rate,
- Animal cells have a low expression rate ,
- Requirement of growth conditions are complex for animal cells/ tissue and
- Animal culture media preparations are complex.

6.5   Products and Applications

Many biotechnological production processes and research programs depend on successful cultivation of animal cells *in vitro*. Production of monoclonal antibodies is one such example, apart from a wide range of biopharmaceutical products produced in genetically engineered mammalian cells. The advantage of using the mammalian cells (i.e. a eukaryotic animal cell and not a prokaryotic bacterial cell) is that they produce the desired product in its correct glycosylated form and that is appropriately folded.

The use of animal cell culture for commercial production and for scientific investigations related to pharmaceutical sciences includes the following:

1. Production of viral vaccines: e.g. cultivation of poliomyelitis virus on Simian monkey kidney cells; cultivation of rabies virus on dog kidney cells.

2. Production of cell surface antigens: the cell surface antigens are used to raise antibodies against themselves which are used for therapeutic or diagnostic purposes E.g. cultivation of surface antigens of human leukemia cell line against which antibodies can be generated.

3. Culturing of immune cells: e.g. to study the effects of immuno-modulators.

4. Production of monoclonal antibodies

5. Production of therapeutic proteins: e.g. growth factors, immuno-modulators

6. Production of α interferon: this involves cultivation of leukocytes and infecting the cultured cells with Sendai virus

7. Production of β-interferon: this involves cultivation of human fibroblasts or recombinant mouse α-cell lines and inducing them to recreate the interferon.

8. Production of human hormones: e.g. human chorionic gonadotropin (HCG) from placental tumor cell line 'Be Wa'.

9. Production of enzymes: e.g. urokinase, plasminogen activator, from continuous line of porcine kidney cells (LLC-PK1)

10. *In vitro* skin cell growth in clinical practice,

11. Investigations related to pharmaceutical sciences: some of these are presented in Table 6.3.

Table 6.3 Scientific investigations related to pharmaceutical sciences involving the use of animal cell culture.

Cell Culture	Investigations
CaCo-2 Cell culture	Estimation of absorption of peptides (cephalosporin, transferrin) across intestinal membrane
Primary culture	Evaluation of peptide (delta sleep inducing peptide, vasopressin) bovine BMC absorption across blood brain barrier
Hepatocytes	Determination of peptide absorption across sinusoidal and cannicular membranes of hepatocytes
Primary fibroblast	Identification and study of hereditary metabolic disorder like cultures acatalasia, galactosemia and maple syrup urine disease
Various cultures	Study of human genetic disorders like glycogen storage disease, xerodermia pigmentation etc.
	Virology: host parasite relation steps
	Neoplasia: understanding the problem of cancer.

6.6   Transgenic Animals

This refers to an animal whose genome has been altered to contain a gene from another animal usually from different species. Any method or means by which foreign DNA is transferred into an animal cell is termed as transfection. The animal which has genetically been altered is referred to as genetically engineered or genetically manipulated (GM).

6.6.1 Methods of gene transfer or transfection

1. ***Transfection of egg cells:*** - The aim here is to introduce the desired whole nuclei, whole chromosome or a fragment of DNA into an egg cell. This is achieved by ex-nucleating the egg cell and then either incubating it with a co-precipitated chromosome or by micro-injecting the DNA/ nucleus into the egg cell.

    The modified egg is then transferred to uterus of a surrogate mother to produce the transgenic animal possessing superior qualities/ triats.

2. ***Lipofection:*** - This is delivery of DNA into cells using liposomes.

3. ***Micro-injection:*** - This involves injecting DNA into the egg's DNA.

4. ***Electroporation:*** - This involves application of high voltage gradient (4000-8000 V/Cm), to a mixture of cell and DNA. This results in the formation of transient pores in the cell membranes through lUgh which DNA can be internalized.

5. ***Retroviral vectors:*** - Recombinant retroviruses carrying the gene of interest splices it into the host's genome. It is especially used to infect 4-16 cell mice embryos.

6. ***Targeted gene transfer:*** - This involves homologous recombination such that the transgene replaces its allele in the genome at the desired position.

7. ***Embryonic stem cell transfer:*** - If cells from two or more very early (pre-implantation) embryos are taken and mixed together and transferred to the surrogate mother then the 'created' animal would have traits/ characteristics of the two or more animals. Such an animal is known as "chimera". (E.g: mythological chimera which had a lion's head, a goat's body, serpent's tail and breathed fire; or the Hindu mythological *'Narsihima Avtar'*). Such chimeras are a reality e.g. the goat/ sheep chimera (called 'geep') and the cow/ buffalo chimera. But, further research on production of such chimeras is banned by the Governments the world over; but quite possibly it is being carried out in secrecy. No one knows!

6.6.2 Applications

Transgenic animals are used for the following three properties:

1. ***Creating animal models for disease:*** - With animal models of known disease, it becomes possible to study how the disease starts and progresses and then to carry out experimentation studies. Transgenic mice are most commonly used to mimic human disease and for screening of potential new drugs. Several genetic diseases like alzheimer's disease, arthritis, hypertension and muscular dystrology have been studied using transgenic mice as a model system.

2. ***Pharming*** i.e. employing transgenic animals for producing pharmaceutically useful proteins: - The gene is spliced onto a promoter signal and a signal sequence so that the foreign protein is expressed in the mammary glands and the protein is thus secreted in the milk. Companies involved in pharming include Pharmaceutical Proteins Ltd and Genzyme. The animals used include sheep and goat (and not the cows due to their larger breeding cycles and small number of off springs) and pigs (due to their short gestation period). The products produced include, casein, lactoglobulin, alpha-1-anti trypsin (for treating emphysema), tissue plasminogen activator, hGH, urokinase, blood clotting factor IX, antithrombin III, cystic fibrosis trans-membrane regulator etc.

   The idea of 'farming' is being investigated using hen's eggs e.g.: for production of antibodies by the company, Transxenogen.

3. ***Farm animal improvement:*** - in terms of:
   - Increased milk production in cattle.
   - Increased growth rate in fish.
   - Improvement of wool production in sheep.

6.7 Cloning

Clone is a collection of genetically identical individuals. Cloning means to make several copies of an individual that are genetically identical. Identical (not fraternal) twins are clones of each other. For cloning an animal, we cannot simply use any portion of the human body and expect it to grow into an individual as in the case with plants using 'cuttings'. The cells that can give rise to a new individual include the fertilized egg and cells from very early (pre-implantation) embryos. Therefore, there are two possible ways to obtain clones.

1. Split an early embryo into maximum number of eight cells (as is the case for natural identical twins!) and transfer them into surrogate mothers to allow these cells developing into identical clones. It is to be noted that if the embryo is split into more than eight cells/ parts, the chances of developing into fetuses of any of the cell/ parts diminishes. Remember 100 *Kauravas*! They simply could not have been produced by splitting the embryo into 100 parts!

2. Injecting the nucleus, taken from another cell, into a fertilized enucleated egg cell. This method is the one that is presently being used for cloning animals like pigs, cattle, goats, mice, cat, from somatic adult cells.

The technique involves removing the nucleus from a fertilized cell (zygote) and then injecting the nucleus of a somatic cell into the egg. Here, also, it is to be noted that the more differentiated the somatic cell is, the harder it will be to 'reprogram' it to allow the development of a whole individual. Therefore, embryonic cells and stem cells are considered to be good candidate for developing clones. Dolly, the typical example of animal cloning, was developed by injecting the nucleus from an adult mammary gland cell into the enucleated cell.

Further, it is the mammals that are difficult to clone. Fish and frogs are relatively easier to clone and their cloning is well established.

As far as human cloning is concerned, it is considered unethical, immoral and impractical (at present). However, therapeutic cloning i.e. cloning of tissues or organs from stem cells from individuals for therapy, so that the organ is not rejected when transplanted into that individual should be allowed. Though, the technology has not developed so far to allow the controlled differentiation of stem cells into the desired organ without letting the stem cell or fertilized egg develop into the whole individual!

7 # MONOCLONAL ANTIBODIES

7.1 Introduction

In 1890, Behring and Kitasato found that rabbits immune to tetanus had 'substances' in their blood capable of destroying the tetanus toxin. The fraction of blood having the activity was called anti-toxin and the substance was identified as immuno-globulin or antibody. In 1891, Behring and Kitasato employed anti-toxin, isolated from rabbits immunized with diphtheria, to treat children suffering from diphtheria. Paul Ehrlich recognized the potential of antibodies and in 1928, he used the term 'magic' bullets for antibodies since they accurately target and neutralize their antigens. He laid down the first theoretical principles of antibody-antigen interactions.

Antibodies or immuno-globulins are serum glycoproteins synthesized and secreted by plasma cells which are derived from β-lymphocytes, stimulated in response to foreign substances called antigens. The production of a specific antibody against an antigen is explained on the basis of clonal selection theory, developed by Macfarlane Burnet, Niels Jerne, David Talmage and Joshna Lederberg in 1957. One of the essential features of this theory is that each antibody producing cell i.e. each B cell makes antibody of a single kind. Therefore, when an antigen is injected in an animal or human, the serum is found to contain a complex mixture of antibodies produced by the stimulation of a number of B cells. Therefore, (in contrast to enzymes) antibodies though having a common specificity are normally heterogeneous because a number of B cells recognizing the antigen are stimulated each generating a clone of plasma cells, each of which produces a different antibody (polyclonal). Thus, if antibodies are to be employed for therapeutic and diagnostic purposes or to elucidate mechanism of biological process, their heterogenicity is to be overcome. Simply this means than, to isolate one B cell and allow its cultivation to form a clone which will synthesize and secrete homogenous antibodies. But the B cells are mortal, they cannot be cultivated indefinitely.

To overcome the problem of mortality, Caesar Milstein and Georges Koehler in the year 1975, came up with the idea of fusing the B cells with the myeloma cells. Myeloma is a malignant (cancerous) disorder of lymphocytes where in the cells divide uncontrollably i.e. they are immortal. Therefore a hybrid of a B cell with a myeloma cell results in the formation of 'hybridoma' which has the property of indefinitely (immortalized), producing the homogenous (monoclonal) antibody as specified by the parent B cell. In 1984, they were awarded the Noble prize for their ingenious technique.

But for the technology to develop and for the monoclonal antibodies to commercialize it has taken almost two decades. As of today, monoclonal antibodies are a hot-cake item accounting for about 25% of the biotech based pharmaceuticals.

7.2   Production of Monoclonal Antibodies

Monoclonal antibodies for human use can be produced by a number of ways. The first monoclonal antibody to hit the market was Orthoclone OKT3 (product of Orthobiotech) in the year 1986. This was produced using the hybridoma technology using the mouse spleenocytes (B cells). Then after, several other monoclonal antibodies have been introduced in the market which are more 'humanized' or are 'human' monoclonal antibodies rather then being 'murine'. So in this section, we'll first consider the production of monoclonal antibodies using hybridoma technology and than go on to discuss other recent methods for the production of monoclonal antibodies.

7.2.1 Hybridoma Technology

Like genetic engineering, hybridoma technology which involves fusion of different types of cells to give a hybrid cell, is part of 'modern' biotechnology. The steps involved in hybridoma technology are:

1. *Immunization:* A mouse is immunized by injecting intradermally or subcutaneously or intravenously, the target antigen mixed sometimes with an adjuvant like aluminum salt to stimulate the immune response.

2. *Spleenectomy*: This means sacrificing the mouse and removal of the spleen which is concentrated source of B cells. The spleen is subjected to mechanical or enzymatic treatment to dissociate into single spleenocytes.

3. *Fusion*: - Spleenocytes are mixed with an immortal cell line i.e. myeloma cells in a ratio of 5:1 to 2:1 and allowed to fuse in the presence of an equal amount (50%w/w) of fusiogenic agent like PEG.

4. *Selection of hybrid:* - When spleenocytes are mixed for fusing with myeloma cells, there can be diversity of possible fusion products (hybrids). The different types of resulting hybrid cells may include spleenocyte-spleenocyte hybrid, myeloma cell-myeloma cell hybrid and spleenocyte-myeloma cell hybrid or no hybrids are formed i.e. single spleenocyte and myeloma cell. In order to select the desired hybrid cells (spleenocyte-myeloma cell) use of a selective medium like the HAT (hypoxanthine,

aminopterine and thymidine) medium was devised by Littlefield. In order to understand the selective nature of the medium, we must first understand as to how the mammalian cells synthesize nucleotides. Nucleotides are biosynthesized by one of the following two different pathways:

- De novo pathway: - In the de novo pathway, nitrogenous bases, purines and pyrimidines are built up from simple precursors like amino acids, formyl tetrahydrofolate and carbon dioxide. A number of enzymes are involved in assembling the nitrogenous bases and then their synthesis to form nucleosides and nucleotides. But one enzyme, dihydrofolate reductase is required for the generation of tetrahydrofolate, which is one of the nitrogenous base precursors. This enzyme is competitively inhibited by aminopterine and amethopterin (latter is also called methotrexate) either of which is present in the HAT medium. Therefore, in case of cells cultured in HAT medium, de novo synthesis does not work. So, it is now the other pathway that the cells have to depend upon if they are to survive.

- Salvage pathway: - In this pathway the nitrogenous bases are not synthesized, instead pre-formed bases coming from the hydrolytic degradation of nucleic acids and nucleotides are used to synthesize the nucleotides. Salvage (which means to save) is simpler and less costlier pathway than de novo pathway. In the HAT medium, hypoxanthine (for synthesis of purine nucleotides) and thymidine (for synthesis of pyrimidine nucleotides) are included. In order to utilize these substrates to form nucleotides, two key enzymes are needed: hypoxanthine-guanine-phosphoribosyl transferase (HGPRT) and thymidine kinase (TK).

HGPRT catalyses the formation of inosinate from hypoxanthine. Inosinate is the precursor from which both (purine nucleotides) adenylate and guanylate are synthesized. TK catalyses the phosphorylation of thymidine (a nucleoside) to thymidilate (a nucleotide).

In mammalian cells there are genes which code for these two enzymes (HGPRT and TK) and if these genes are functional in a cell then the cell can survive in the HAT medium using the salvage pathway. Now as described under animal culture, normal animal adult cells are not able to survive *in vitro* for long period of time probably due to aging or senescence (i.e. programmed natural cell death). Thus B cells cultured *in vitro* in HAT medium, though possessing functional HGPRT and TK genes, die out. It is the cancerous cells (myeloma cells) that can be cultured indefinitely. Here, in case of hybridoma technology, myeloma cells that have non-functional or are deficient in HGPRT and/or TK genes are used. Thus, myeloma cells also die out in the medium. It is only the fusion product i.e. hybrid of myeloma (deficient in $HGPRT^-/TK^-$ gene(s)) and B cells (having functional $HGPRT^+$ and TK^+ genes) that are able to survive and grow indefinitely (a property of myeloma cell) and produce the antibody (a property of B cells).

Therefore, the medium allows selective growth of hybridoma cells while the original cells die out.

5. ***Cloning by dilution:*** - The population of cells proliferating in the HAT medium consists of hybridoma cells, all of which secrete antibodies. The aim now is to screen the hybridoma cells, to determine which ones produce antibody having the desired specificity. In order to be able to do this the cells in cultures need to be diluted and a small amount of the culture is transferred into 96-well plastic plates. Ideally, the dilution is to be such that only one positive cell is found per well. The cell in the well is allowed to grow repeatedly to form a clone.

6. ***Screening:*** - The next step is to screen the cells producing a 'good' antibody of desired specificity i.e. an antibody that binds tightly to an antigen, having affinity of 10^9 or more, is highly specific for the particular antigen and is of right class and subclass. The screening is performed in mass using 96-well plastic culture plates to the bottom of which the desired antigen is bound. To identify the desired antibody producing cell, methods like ELISA, RIA, FIA are employed. The cells producing the desired antibody are termed positive cells. The positive cells can be grown for production of antibodies or alternatively can be frozen and stored for a long period of time.

7. ***Production:*** - Single positive cells are then grown either using tissue culture or by injecting into mice. The cells injected into mice form an ascetic tumor and secrete the antibody in ascetic fluid surrounding the lungs and also in plasma. The amount of antibodies produced using mice is fairly large (1-25 mg/ml) than the amount using tissue culture methods (10-100 μg/ml). Also the hybridoma cells can be cultured in immobilized cell reactors like the hollow fiber reactor and also in the traditional suspended cell fermentors. e.g.: A 1000 liter air lift fermentor produces 100 gm of antibody in two minutes time.

Other methods explored for the production of monoclonal antibodies include genetic engineering of CHO cells; pharming i.e. production in animal milk; production in eggs, plants, bacteria and viruses etc. Some of the novel techniques used for the production of monoclonal antibodies shall be considered in the following section. To appreciate the need for evolving novel techniques and production of humanized and human monoclonal antibodies, it will be a good idea to first understand the structure of antibodies and the reason for the presence of huge diversity of antibodies.

7.3   Structure of Antibodies

Chemically antibodies are glycoproteins i.e. proteins with attached carbohydrates and are termed immunoglobulins (Ig) since they take part in immune response. All antibody molecules have two basic functions:

1. Binding with the antigen and
2. Participation in the effector functions (like neutralization, agglutination, precipitation, lysis, complement-fixation or opsonization).

These two properties reside in two distinct fragments of the antibodies. The two fragments can be separated by cleavage of the antibody molecule by treatment with papain, a proteolytic enzyme. The two fragments of the antibodies are:

- antigen-binding fragment abbreviated as Fab and
- crystallizable fragment abbreviated as Fc (see Figure 7.1)

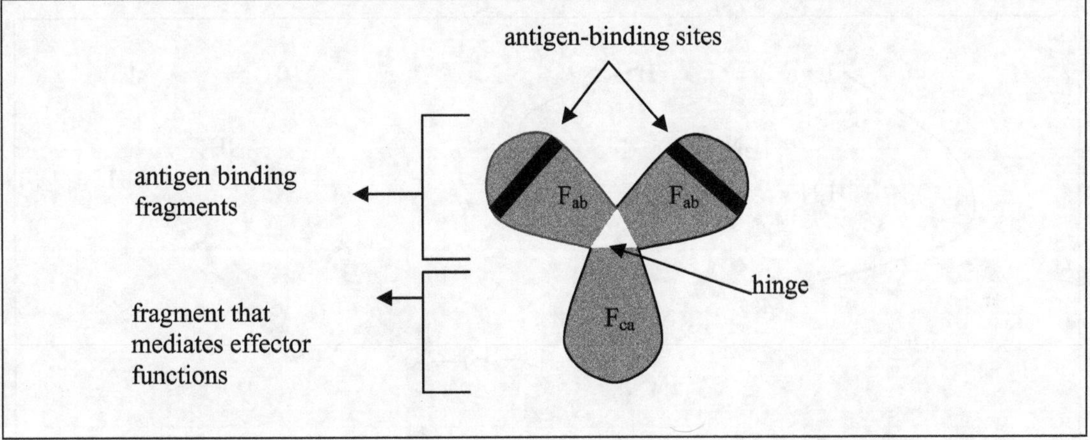

Figure. 7.1 Typical structure of an antibody depicting the two distinct fragments.

The antibody molecules consist of: (refer Figure7.2 as you read along)

1. Two large and small pairs of polypeptide chains
2. Large pair is composed of two heavy (H) chains i.e. each heavy chain having a molecular weight of 50 Kdalton
3. Small pair is composed of two light (L) chains i.e. each light chain having a molecular weight of 25 Kdalton.
4. Each of L & H chains are bonded together by a disulphide bond.
5. The H chains are bonded together by at least one disulphide bond.
6. Fab consists of entire L chains and the amino terminal halves of the H chains.
7. Fc consists of carboxyl terminal halves of both the H chains.
8. The L chains are made up of 214 amino acid residues.
9. The H chains are made up of 446 amino acid residues.
10. The L chains consist of variable region (from 1 to 108 amino acid residues) and a constant region (from 109 to 214 amino acid residues).
11. The H chain consists of variable region (from 1 to 108 amino acid residues) and a constant region (from 109 to 446 amino acid residues).
12. There are three hyper variable segments (generally termed complementarity-determining regions CDR) present in variable regions of both the L & H chains.

13. It is these six CDRs of the antibody that determine antibody specificity and affinity and that actually make contact with the epitope i.e. the specific region of antigen contacted by the antibody.

14. The other parts of the variable regions that remain relatively unchanged serve as frame work for proper orientation of CDRs, so that a close contact is made possible with the epitope.

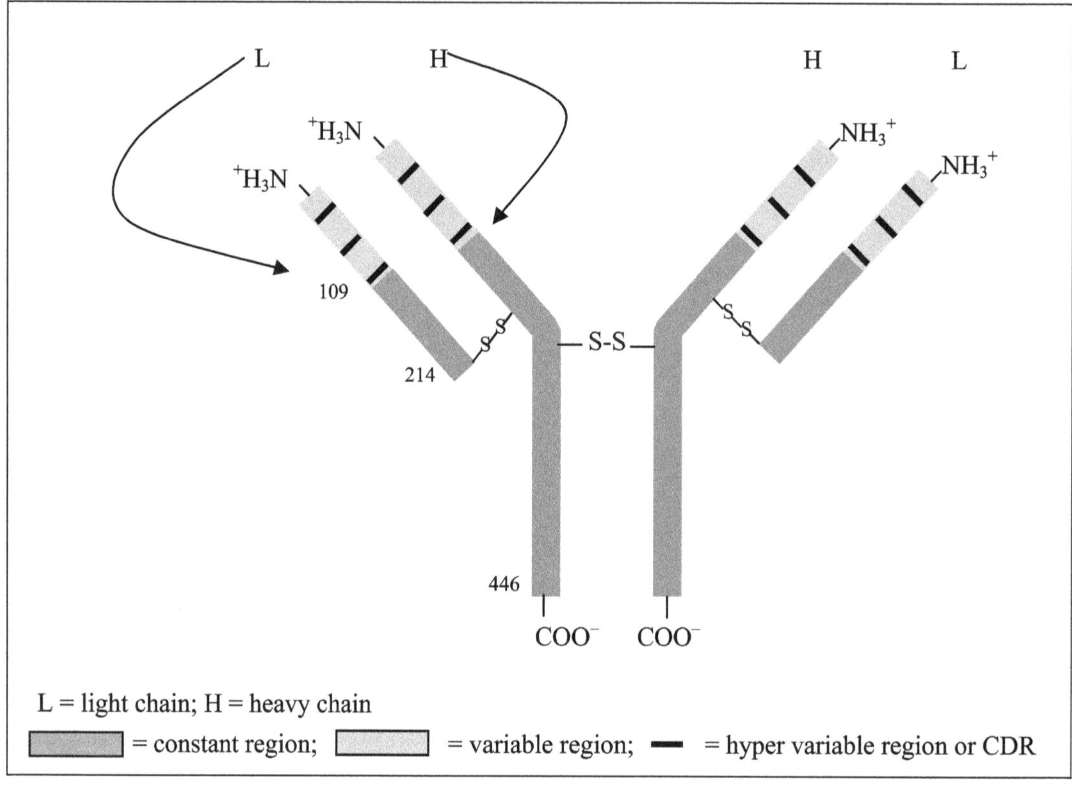

Figure. 7.2 Structure of antibody.

7.4   Novel Approaches for Production of Monoclonal Antibodies

Orthoclone OKT3, used in the prevention of renal transplant rejection, was the first monoclonal antibody that was launched in the year 1986. Since then about a dozen other Monoclonal antibodies have been launched and three scores are under clinical trials. Orhtoclone OKT3, a murine monoclonal antibody, is produced using the hybridoma technology; while almost all the other monoclonal antibodies for therapeutic use (not for diagnostic use) in the market are produced using one of the novel approaches. Murine monoclonal antibodies, i.e. antibodies produced in mouse, illicit severe immune reactions in human patients. The reason is obvious, immunoglobulins are proteinaceous in nature and when mouse antibodies are introduced in humans, the cells of the immune system recognize them as foreign (antigen) and are stimulated to produce antibodies, termed human anti-mouse antibodies (HAMA). Of course it takes time for the immune response to develop HAMA (8-12 days) and thus monoclonal antibodies of murine origin cannot be employed for therapeutic purposes for more than ten days. Thus, murine monoclonal antibodies are unsuitable and therefore there is a need to develop humanized or even better human monoclonal antibodies, the approaches to the development of which are considered in this section.

7.4.1 Chimeric Monoclonal Antibodies

As seen in the previous section, an antibody has various regions (V_H, V_L, C_H and C_L) and there are genes which encode for the various regions (V and H genes) and other joining regions (D and J genes). In chimeric antibodies, part of the genes encoding for the antibody are derived from mouse and partly from human, which results in the formation of antibodies that have properties in part like murine antibodies and in part like human antibodies.

Chimeric antibodies are made by creating chimeric genes wherein the genes encoding for the variable regions of the monoclonal antibody are obtained from desired (selected and screened) hybridoma (murine) and genes encoding for constant regions are derived from human B-cells. The created chimeric gene is then introduced into Chinese hamster ovary (CHO) cells wherein chimeric monoclonal antibody is expressed. The antigen specificity (due to Fv) is identical to murine antibodies, while the effector function (due to Fc) and *in vivo* half-life are identical to human antibodies. Thus, these monoclonal antibodies are about 75% human (not fully human). These are less immunogenic (i.e. to be recognized as foreign to illicit an immune response) than murine monoclonal antibodies but still have the potential to stimulate the production of human anti-chimeric antibodies (HACA). Therefore, development of monoclonal antibodies has still gone a step further to produce humanized (close to the human) monoclonal antibodies.

7.4.2 Humanized Monoclonal Antibodies

Considering the structure of an antibody, it will be readily seen that the region in the Fv that is responsible for actually making the contact with the epitope on an antigen is located in the CDR, while the adjoining region provides for the framework: There are genes which encode for CDRs. Thus, a humanized antibody gene is constructed by utilizing only these sequences from the murine antibody gene that encodes for the hyper variable region and utilizing rest of the gene sequences (for framework and Fc) from the (human) immunoglobulin G (IgG1)sequence. The resulting antibody is 90% human and data concerning its immunogenicity are lacking at present.

The development can still go further, since the ultimate aim is still to produce fully non immunogenic, human antibodies.

7.4.3 Human Monoclonal Antibodies

These can be produced by one of the two approaches described below.

7.4.3.1 *Phage display*

In this approach about 1-2 millions of genes are constructed encoding for the variable regions of both the light and heavy chains with a predominant linker between them, so as to express them (V_H & V_L fragments) as part of a single polypeptide chain. These are called as mini antibodies or single chain variable fragments (ScFv). The V_L-linker - V_H or the ScFv genes are spliced upstream of gP3 gene in DNA of phage such as M13. gP3 is one of the constituent proteins of the phage outer coat. Such phages containing the recombinant ScFv-gP3 gene are allowed to replicate in bacterial cell culture where they (M13 bacteriophages) express the antibody fragments as part of the outer protein coat. This results in the generation of what is called as phage display antibody library. Now such a phage that has appropriate antibody on its surface and binds most strongly to the antigen needs to be selected from the library. Such a selected phage also has the gene that can generate human antibodies.

7.4.3.2   *Human antibody mouse*

Complete human antibodies can be made using mice in which genes encoding for heavy and light chains have been inactivated (said to be "knocked out") by homologous DNA recombination. This means that such genetically engineered mice lack B-cells and will not produce antibodies. In such mice, DNA segment encoding for human heavy and light chains of antibodies are inserted. These 'Human monoclonal antibody' mice are then immunized with an antigen to stimulate the production of antibodies. Spleenocytes of the immunized mice are collected and allowed to fuse with myeloma cells to form hybridomas. The hybridomas are screened and hybridoma producing the best desired monoclonal antibody is selected for the production of fully human monoclonal antibody.

The various types of monoclonal antibodies are summarized in Table 7.1.

Table 7.1 Various types of monoclonal antibodies.

Antibody Type↓/Fragment→	Variable Framework	CDR	Constant	Example
Murine	M	M	M	Muromomab (OKT3)
Chimeric	M	M	H	Basiiximab
Humanized	M	H	H	Transtuzumab
Human	H	H	H	-

Abbreviations: M- murine;   H – human

7.5   Nomenclature

For naming monoclonal antibodies, US Adopted Name (USAN) Council has developed the following guidelines:

1. The suffix *–mab* is used for monoclonal antibodies and fragments
2. The source of the antibody is designated within the name. The following are approved abbreviations for the source:

a = rat	u = human	e = hamster	xi = chimera
i = primate	zu = humanized	o = mouse	

3. The general disease state i.e. subclass is included in the nonproprietary (generic) name by inclusion of a code syllable approved by the USAN Council as follows:

viral	= vir	tumour of mammary gland	= mar
bacterial	= bac	tumour of testis	= got
immunomodulator	= lim	tumour of ovary	= gov
infectious lesions	= les	tumour of prostate	= pr(o)
tumour of colon	= col	tumours-miscellaneous	= tum
tumour-melanoma	= mel	cardiovascular	= cir

4. A distinct and compatible syllable is selected as the beginning *prefix* in order to create unique name.
5. Thus the order for combining the various key elements is as follows:

    Unique prefix followed by disease state/target followed by source of antibody followed by suffix mab.

6.   If the product is conjugated to another chemical or is radiolabelled, recognition is made by the addition of a second word.

7.6   Applications

Broadly speaking, the applications of monoclonal antibodies can be grouped under the following heads:

1. Therapeutic
2. Diagnostic
3. Investigational and analytical
4. Drug targeting
5. Abzymes (catalytic antibodies)

The specific uses of the commercially available monoclonal antibodies are listed in Table 7.2.

Table 7.2 Applications of commercially available monoclonal antibodies.

Generic name	Brand name	Company	Use
Therapeutic			
Abciximab	ReoPro	Centocor	prevention of blood clots
Basiliximab	Simulect	Novartis	prophylaxis of acute organ (renal) transplantation rejection
Daclizumab	Zenepax	Hoffman La-Roche	prevention of acute kidney transplant rejection
Endrecolomab	Panorex	Centocor Glaxo	treatment of colorectal cancer
Gemtuzumab	Mylotarg	Celltech	treatment of acute myeloid leukemia
Infliximab	Remicade	Centocor	treatment of Crohn's disease; rheumatoid arthritis
Muromomab	Orthoclone OKT3	Ortho Biotech	reversal of acute kidney transplant rejection
Palivizumab	Synagis	MedImmune Abbott	prophylaxis of lower tract disease caused by syncytial virus
Rituximab	Rituxan Mabthera	Genentech Hoffman La-Roche	treatment of Non-Hodgkin's lymphoma
Transtuzumab	Herceptin	Genentech	treatment of Metastatic breast cancer
Diagnostic			
			detection of-
Capromab	ProstaScint	-	prostate adenocarcinoma
Imiciromab pentetate	MyoScint	-	myocardial infarction
Nofetumomab	Verluma	-	small-cell lung cancer
Satumomab pendetide	OncoScint	-	colorectal and ovarian cancer
Technitium -99-arcitumomab	CEA-Scan	-	metastatic colorectalcancer

8 # MICROBIAL TRANSFORMATION

8.1 Introduction

Microbial transformation, also called as biological transformation or bioconversion, is simply the conversion of one compound into another brought about by living microorganisms. Biotransformation, on the other hand, is the conversion brought about by enzymes in a living organism. *Bioconversion* (as in traditional organic chemistry in most cases) means to synthesize a complex compound such as steroids, prostaglandins, antibiotics etc. while *Biotransformation* refers to a detoxification process such as metabolism of drugs.

Why is it essential to restore to bioconversion for synthesis of drugs? Can synthesis of compounds not completely be carried out the *invitro* in the laboratory? The answer to the latter question is yes (and at times no) we can (and cannot). But the synthetic route may involve in some cases scores (20s) of steps to be carried out under extreme conditions requiring a long time. For instance, it takes 37 steps to convert ox bile salt (steroidal nucleus) to cortisone. Further, the enzymes required are unstable under the laboratory conditions and thus breakdown. The microbes on the other hand, continuously make the required enzymes. Answer to the first question is that microbes are able to bring about the regioselective (position specific) and steriospecific (enantiomer specific) addition, removal or modification of the functional groups on a complex molecule without having to protect the other labile chemical groups (i.e. microbes are reaction specific). Further, the microbes bring about the conversion of a compound into a structurally related, functionally more valuable compound at moderate conditions of temperature (20°C-40°C), pressure (0-34kPa), pH (around neutrality) in aqueous media without the requirement of heavy metal (which are pollutant/contaminant in the product) catalysts.

For bioconversion, microbial cells (microbial transformation), plant cells, animal cells or the required enzymes are isolated from organisms and immobilized. The immobilized cells/enzymes can be employed to convert a substrate (compound) in the fermentation medium into the desired compound. However, most commonly, it is the microbes that are employed because of their rapid growth and high rates of metabolism in comparison to the whole plant or animal cells. It is to be noted that for a microbial strain to bring about the bioconversion effectively, the strain has to be genetically optimized. This is because microbes tend to utilize the substrate (here it is the compound) effectively for their growth and multiplication rather than convert it to the desired compound. Or if they convert the substrate (compound) to the desired compound, then it is done inefficiently.

Microbes, bringing about bioconversions, catalyze a wide range of chemical reactions such as:

- Oxidation
- Decarboxylation
- Amination
- Deamination
- Isomerization

Among the well established products of microbial transformation include:

- Gluconic acid
- Acetic acid
- Steroids
- Prostaglandins

8.2 Applications

In this section, the importance of microbes in bringing about bioconversion of compounds to produce the desired, high value products shall be considered. The products include:

8.2.1 Acetic Acid

This production of acetic acid involves bioconversion (oxidation) of ethanol through acetaldehyde to acetic acid brought about by species of *Acetobacter*.

8.2.2 Gluconic acid

This production of gluconic acid involves one step enzymatic bioconversion (glucose oxidase) of glucose to gluconic acid brought about by *Aspergillus niger*.

8.2.3 Steroids

The production of steroids such as *corticosteroids*: cortisone, prednisone, prednisolone, hydrocortisone, dexamethasone; *androgens*: testosterone; *estrogens*: estradiol; etc. involves bioconversion of the relatively inexpensive starting materials. The starting materials include:

- Masterols and sitosterols- obtained as by-products in the production of soybean oil,
- Diosgenin- obtained from the roots of Mexican barbisco plant, *Dioscorea compopsitae* and the Indian *gokhru* and
- Deoxycholic acid- obtainaed from animal (ox) bile.

Transformations of the steroids can be brought about by employing moulds, bacteria, actinomycetes, yeast, protozoa, algae and even cell free enzyme preparations. However, most commonly, it is the fungi and bacteria that are employed commercially. The structure of the steroidal nucleus is shown in Figure 8.1. The reactions that can be catalyzed by microbes to transform the steroids include:

1. Hydroxylation: Both α and β on almost every position
2. Dehydrogenation: resulting in the insertion of a double bond e.g. Δ^1, Δ^4
3. Hydrolysis: of ester to alcohol
4. Isomerization: Δ^5 to Δ^4
5. Oxidation: of alcohol to ketone e.g. 3β-OH to 11-CO; 11β-OH to 11-CO; 17β-OH to 11-CO

6. Cleavage of side chain at C-17
7. Expansion of Ring D
8. Lactone formation
9. Oxidation

Thousands of transformed steroidal compounds can be produced based on the above mentioned wide ranging reactions. However, the position of key interest in steroidal nucleus is C-11. It is at this position that an oxygen atom is very much required for biological activity (anti-inflammatory). Chemically introducing the oxygen atom at C-11 is a pretty tough job. It is here that the fungal species are used commercially to bring about the transformations. Some of the noted microbial transformations involved in the production of steroids are presented in Table 8.1.

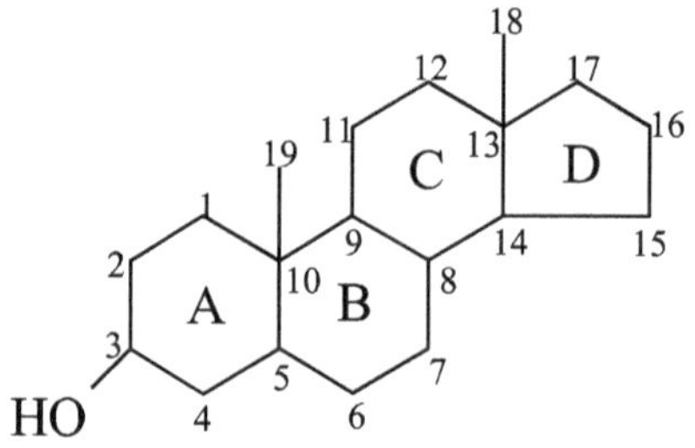

Figure 8.1 Structure of steroidal nucleus.

8.2.4 Prostaglandins (PGS)

Eicosanoids, (derived from eikosi which in Greek means 20) are molecules containing 20-carbon atoms. These include prostaglandins, prostacylin, thrombaxane and leukotrienes.

Prostaglandins (PGs) are 20-carbon fatty acids containing a 5-carbon ring (cyclopentane) with α- and ω-side chains. The major classes of prostaglandins that are known are classified and designated as PGA through PGI. A number in subscript in the abbreviation denotes the number of double bonds outside the ring (for e.g. PGA_2). Thus, the biosynthetic precursor of eicosanoids, including PGs, is the 20-carbon fatty acid: arachidonic acid (Figure 8.2). Microbes can be employed to produce PGs utilizing unsaturated and polyunsaturated fatty acids (PUFA) as substrates. The advantage of employing microbes (over chemical synthesis) is of course that their reactions tend to be steriospecific, producing the desired enantiomer. PGs possess at least three centers of chirality! Some of the noted microbial transformations involved in the production of PGs are presented in Table 8.2.

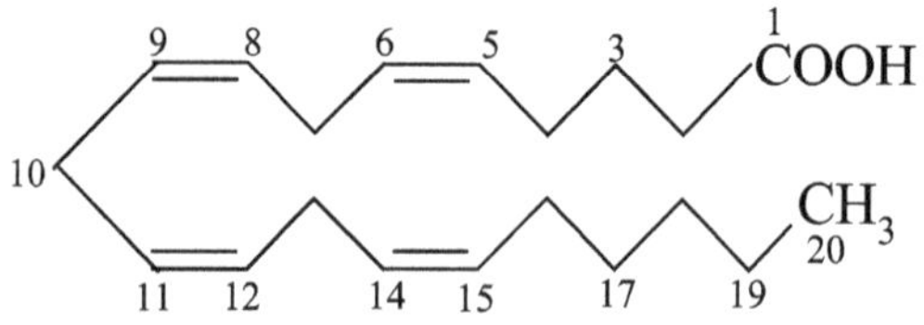

Figure 8.2 Arachidonic acid.

8.2.5 Miscellaneous Compounds

Some more noted and commercially important transformations brought about by the microbes are complied in Table 8.3.

A lot of other microbe mediated transformative reactions have been reported where the transformed product has not found to be of much importance in therapy. However, these reactions of historic importance led to the development of systematic study of microbial transformations.

A few of these transformative reactions include:

1. ***Transformative reactions of historic/academic importance:***
   i. Oxidations:

      D-sorbitol to L-sorbose

      Glucose to 2-ketogluconic acid

      Isopropanol to acetone

      Polyhydric alcohols to corresponding sugars
   ii. Hydrolysis

      Di-, tri- saccharides to constituent monosaccharides

      Tannins to gallic acid
   iii. Reduction

      Fructose to mannitol

      Malic acid to succinic acid
   iv. Racemate resolution of

      Glyceric acid

      Lactic acid

      Tartaric acid

2. ***Transformations involving antitumour drugs:***
   i. Reductive cleavage of

      Aminoglycosides
   ii. Reduction of

      keto groups in anthracyclines
   iii. Hydrolysis of

      amine moiety of bleomycin
   iv. Regioselective hydroxylation in

      Withaferin A; aromatic rings of acromycin; vinblastine
   v. N-demethylation in

      D-tetrandrine; Vindoline
   vi. Dimerization of

      Vindoline
   vii. Opening of

      Quinine ring in lapchol

Table 8.1 Some of the noted microbial transformations involved in the production of steroids.

Starting material	*Microorganism(s)* **Reaction(s) catalyzed**	Product of transformation	Further, chemically modified into
Progesterone	i. *Aspergillus ochraceus* ii. *Rhizopus arrhizus* iii.*R. nigricsns* 11α-hydroxylation	11α-hydroxy progesterone	Cortisone
17-acetate of Reichstein's substance S	i. *Curvularia luvata* ii. *Cunninghamella blakesleena* 11β-hydroxylation	Hydrocortisone 17-acetate	Hydrocortisone
9α-Flurohydrocortisone	i. *Streptomyces sp.* 16α-hydroxylation	9α-Fluro-16α-hydroxyhydrocortisone	Triamcinolone
11-Deoxycorticol	i. *Curuularia lunata* ii. *Curuularia blakesleena* 11β-hydroxylation	Cortisol (Compound F)	-
11-Deoxycortisol (Reichstein's Substance S)	i. *Pseudomonas sps.* 11β, 17α-dihydroxylation	Cortisol	-
1,2 saturated compounds: Cortisone/cortisol	i. *Arthrobacter (Cornybacterium) simplex* ii. *Bacillus sphacricus* iii. *Bacterium cyclooxydans* iv. *Cylindrocarpon rodiciola* v. *Fusarium caucasicum* vi. *Fusarium solani* vii. *Septomyxa ofinis* viii. *Streptomyces lavendulae* x. *Streptomyces trodie*	Prednisone Prednisolone Triamcinolone 6-methyl prednisolone Dexamethasone	-

Table 8.1 Contd….

Starting material	*Microorganism(s)* Reaction(s) catalyzed	Product of transformation	Further, chemically modified into
	1,2-dehydrogenation		
Prednisone Prednisolone	i. *Bacillus megaterium* Hydrogenation of Δ^1	Cortisone Cortisol	-
4-androstane 3, 17-dione	i. *Bacillus putrificeus* Hydrogenation of Δ^4	Androstane 3, 17-dione	-
Progesterone-I	i. *Rhizopus nigraricus* Hydrogenation	11α-hydroxy-17α-progesterone	-
Tiolone triacetate	i. *Flavobacterium dehydrogenans* Coupled reaction: a. Hydrolysis: Ester to alcohol b. Oxidation: 3β-OH-5-ene to 3-keto-4-ene	17-acetateof Reichstein's substance S	Hydrocortisone
4-androsterene-3-17dione	- Reduction of 17-keto group	Testosterone	-
Secodione	- Steriospecific reduction of keto group yielding desired enantiomer	Secolone	D-Norgesterol
9β(11)-dehydro compounds (I)	i. *Curvularia lunata* ii. *Curvularia blakesleena* 9β, 11-epoxidation	11β-hydroxyl compounds	-

Table 8.1 Contd...

Starting material	Microorganism(s) Reaction(s) catalyzed	Product of transformation	Further, chemically modified into
14(15)-dehydro compounds (II)	i. *Cuvularia lunata* ii. *Cuvularia blakesleena* 14α, 15-epoxidation	14α-hydroxyl compounds	-
19-hydroxycholesterol 19-hydroxy-β-sitosterol 19-hydroxylated stigmasterol 19 nor (without methyl group) Stigmasterol Compesterol Siboester	i. *Nocardia restricta* Coupled reactions- a. C-1 dehydrogenation (Δ^1): resulting in spontaneous ring A aromatization b. side chain cleavage: resulting in formation of 17-CO from 17β-OH	19hydroxy-4-androstene, 3, 17-dione	Esterone
,,	i. *Mycobacterium sps.* a. Ring A aromatization b. Side chain cleavage	1, 4-androsta-diene, 3, 17-dione	Esterone

Table 8.2 Some noted microbial transformations involved in production of prostaglandins.

Starting material	Microorganism(s) Reaction(s) catalyzed	Product of transformation	Further, chemically modified into
Arachidonic acid	*Ophibolus graminis* Hydroxylation at C-18 & 19 enzymes into PGE	corresponding hydroxy compounds	keto compounds and further cyclized using animal
2-(6-methoxycarbonyl hexyl) cyclopentane-1, 3, 4 trione	*Diplodascus uninucleatus* steriospecific reduction	corresponding 4 (R)-alcohol	cyclopentyl synthon
2-(6-carboxy hexyl) cyclopent-2-en-1-one	*Aspergillus niger* ATCC 9142	corresponding alcohol	cyclopentyl synthon
cyclopentyl synthon	*Penicillium decumbens* introduction of ω-side chain (i.e. (+)-3(s)-indocot-1-en-3-ol)	corresponding alcohol	-
corresponding alcohol	*Rhizopus oryzae*	PGE_1	-
corresponding 15-keto PGE	*Flavobacterium sp.* *Pseudomonas sp* *Rhodotorula glutinis* *Flavobacterium sp.* Reduction of 15 keto group leading to $\Delta^{8(12)}$-15-OH PGE	racemate mixture of corresponding 11, 15 diol PGEs	-
-	*Kloeckera jensenii* Enatioselective reduction of conjugated keto group	Sulprostone	-

Table 8.2 Contd…

Starting material	Microorganism(s) Reaction(s) catalyzed	Product of transformation	Further, chemically modified into
3, 5-cis-diacetate cyclopent-1-ene	*Baker's yeast* saponification *Bacillus subtilis* saponification	3(R)-acetoxy-5(R)-hydroxy cyclopent-1-ene 3(S)-acetoxy-5(S)-hydroxy cyclopent-1-ene	Lactone
9, 15-diketo-11-deoxyprostanoic acid	*Saccharomyces cerevisiae* ATCC 4125 15-keto reduction	corresponding 15-OH compopund	-
9, 11-diketo-15-deoxyprostanoic acid	*Trechispora brinkmanii* 11-keto reduction	corresponding 11-OH compopund	-
Natural PGE	*Baker's yeast* Reduction of 9-keto group	PGF$_{1\alpha}$	-
Natural PGE$_2$	*Baker's yeast* Reduction of 9-keto group	PGF$_{2\alpha}$	-
PGA$_2$	*Cephalosporium sp.* *Cephalosporium blakesleeana* *Dactylium dendroides*	15 hydroxy-9 oxo prosta -5, 13 dienoic acid	-
Natural & synthetic PGs	*Peicillium sp.* *Mycobacterium rhodochrons* β oxidation	corresponding tetranor, dinor structures	-
Various Natural & synthetic PGs	*Streptomyces sp.* *Microascus trigonosporous*	*corresponding* hydroxylated structures	-

9 # CELL IMMOBILIZATION

CHAPTER CONTENTS AT A GLANCE

9.1   Introduction

Immobilization of cells is the restriction of cell motility, be it plant, animal or microbial cell, within a defined space. Just like enzymes, cells are also immobilized for the same advantageous reasons.

9.2 Advantages of cell immobilization

The advantages of cell immobilization include:

1. easy separation from the culture medium and product
2. continuous operation of fermentor
3. provision of conditions which are conductive for cell differentiation and cell-cell communication
4. production of secondary metabolites in high yields
5. production of cells against disruptive shear forces
6. high cell concentration
7. cell reusability
8. elimination of costly cell recovery processes, cell recycle and washout problems
9. provision of favorable micro-environmental conditions such as: nutrient-product gradient, pH gradient etc.
10. improved genetic stability

However, it is to be noted that immobilization of cells and the advantages mentioned above are applicable only to those cells which excrete the desired product out of the cells. Also, it is important and interesting to note that the live immobilized cells would grow and evolve gases; this can result in mechanical disruption of the immobilizing matrix and lead to leakage of cells.

9.3   Methods of cell immobilization

Immobilization of cells may be either active or passive; as discussed below

9.3.1 Active Immobilization of cells: This is based on the following techniques:

1. Adsorption: on inert support materials such as-
    i.   inorganic materials: e.g. glass, silica, alumina, ceramics
    ii. organic materials: e.g. gelatin, chitosan, activated carbon, wood chips

iii. ion-exchange resins: e.g. DEAE-Sephadex, CMC, Sepharose

iv. polymeric materials: e.g. polypropylene

v. micro-carriers: e.g. dextran beads coated with gelatin, polyacrylamide beads, polystyrene beads, hollow glass beads, cylindrical cellulose beads, polylysine-sterilized-fluorocarbon droplets (micro-carriers, especially dextran based, are used to immobilize anchorage dependent animal cells)

Adsorption may be physical or chemical, depending upon the surface properties of both the support material and the cells. It may involve electrostatic forces (in case of positively charged surfaces: ion-exchange resins, gelatin), covalent binding, hydrogen bonding (polymeric support), Van der Waal's forces.

Advantages of adsorption as a method of cell immobilization include:

- high cell loadings
- adsorption capacity ranging from 2mg/g for porous silica to 250mg/g for wood chips or 10^8 to 10^9 cells/g for porous glass
- simple, inexpensive, most widely used
- direct contact

However, the major disadvantage is that the binding forces are rather weak which lead to washing out (desorption) of cells on application of slight shear.

2. Covalent binding: This is not the preferred method for cell immobilization (unlike enzyme immobilization where it is the preferred method). This is because the surface has to be chemically pretreated for covalent binding with either coupling agents such as glutraldehyde, carbodimide or reactive groups. Both these treatments happen to be toxic to the cells. However, inorganic carriers, such as oxides of titanium and zirconium, which provide compatible functional groups for covalent binding, have been developed. The advantage of this method is that stronger binding forces are involved which ensure stable binding. However, the disadvantage of the method is that the newly grown cells would be lost/ washed out.

3. Cross linking: This is of two types-

i. Direct cross linking: This is brought about by treatment of cells with glutraldehyde resulting in formation of insoluble aggregate. However, glutraldehyde has an adverse effect on the cells' metabolic activity.

ii. Physical cross linking: This is brought about by using polyelectrolytes, polymers (chitosan etc.), and salts (calcium chloride, aluminum hydroxide, ferric chloride etc.)

4. Matrix entrapment: This involves embedding the cells in materials like:

i. Polymers : e.g. agar, alginate, κ-carrageenan, polyacrylic amide, chitosan, gelatin, collagen etc.

ii. Metal screens

iii. Others: e.g. polyurethane, silica gel, polystyrene, cellulose triacetate etc.

Typically, entrapped cell concentration ranges from 10^9 to 10^{11} cells/ml. Most commonly for matrix entrapment of cells, polymer beads are prepared. Bead formation can be achieved by employing one of the following techniques:

i. *Gelation of polymers*: This involves lowering the temperature of a solution (gelatin, agar) till it solidifies. If cell suspension has been added to the polymer solution, then the cells would get entrapped in the formed matrix on solidification.

ii. *Precipitation of polymers*: This involves alteration of pH or solvent of a polymer solution so that the polymer precipitates out of solution. If cell suspension has been added to the polymer solution then the cells would get entrapped in the formed matrix on precipitation. However, it is to be noted that solvent contact may cause cell inactivation or even death.

iii. *Ion-exchange gelation:* This involves solidification when polyelectrolytes react with salt solution to form a solid gel which entraps the cells. The most noted example is the formation of calcium-alginate beads. Other interactions include, aluminum- alginate, calcium-aluminum-CMC, magnesium-pectinate, potassium-κ-carrageenan, chitosan-polyphosphate.

iv. *Polycondensation*: This involves the formation of chemically and mechanically stable polymeric networks via covalent binding involving severe reaction conditions i.e. extreme of temperature, pH etc. Examples of plycondensation polymers include epoxy resins, polyurethane, silica gel, gelatin-glutraldehyde, albumin-glutraldehyde, collagen-glutraldehyde etc.

v. *Polymerization*: This involves cross linking of monomers such as acrylamide, methacrylamide, 2-hydroxyethyl methacrylate to form polymers. A divinyl compound such as methylene bis-acrylamide is most commonly used to initiate cross polymerization. Here also, if a suspension of cells is mixed in the polymerizing solution, then the cells shall be embedded as polymerized beads. The disadvantage with the beads is the leakage of cells due to cell division occurring within the beads.

5. Entrapment within porous matrix: This involves, as the first step, diffusion of cells into preformed porous matrices such as brick, cordierite or porous glass, followed by their growth which subsequently results in the cells getting entrapped (just the same way as a python gets trapped behind bars after swallowing up bait!!!).

The advantages of this method include increased resistance to disintegration in packed beads or stirred vessels (shear); support materials are non-harmful to the cells. The obvious disadvantage is that high cell entrapment cannot be achieved due to limited pore volume.

6. Encapsulation in microcapsules: This involves formation of spherical particles entrapping the cells within semipermeable membrane. Polymers used for encapsulation include: nylon, collodion, polystyrene, acrylate, polylysine-alginate hydrogel, cellulose acetate-ethyl cellulose, polyesters, etc. In this case, the concentration of cells entrapped per unit volume is still higher than that in case of beads.

7. Macroscopic membrane or hollow fibre reactor: This is a device consisting of a shell and tubes. The tubes are made of semipermeable membrane through which nutrients are allowed to flow and pass across the membrane to provide nutrition to the cells. For details refer chapter on 'Animal cell culture'.

9.3.2 Passive Immobilization: Biological Film

This is a form immobilization of microbes that is brought about by microbes themselves in their natural course of growth. This involves growing of microbes on a surface over a polymeric material which is produced by the microbes themselves. This growth of microbes is referred to as biofilm. In a biofilm, a number of different microbes (consortium or community or mixed population) tend to grow. Some of these microbes have the characteristic property of corroding the surface (biocorrosion); some produce a polymer such as mucopolysaccharide. This results in formation of a rough and sticky surface, on which a community of microbes gets 'immobilized'.

9.4   Applications

Some examples of immobilized cells with their applications and support materials, are presented in brief in Table 9.1.

Table 9.1 Some examples of immobilized cells: applications and support materials.

Cell	Support material	Application/ converts
Cell immobilization by entrapment		
Acetobacter sps.	Calcium -alginate	Glucose to gluconic acid
Catharanthus roseus	Polyurethane	Isocitrate dehydrogenase activity
Escherichia aerogenes	κ-Carrageenan	Glucose to 2,3-butanediol
Escherichia coli	κ-Carrageenan	Fumaric acid to aspartic acid
Morinda citrifolia	Calcium -alginate	Anthraquinone formation
Nocardia rhodocrous	Polyurethane	Conversion of testosterone
Rhodotorula minuta	Polyurethane	Methyl succinate to menthol
Saccharomyces cerevisiae	κ-Carrageenan/ Polyacrylamide	Glucose to ethanol
Trichoderma reesei	κ-Carrageenan	Cellulose production
Cell immobilization by surface attachment		
Animal cells	DEAE-Sephadex	Production of hormones
Clostridium acetobutylicum	Ion-exchange resin	Glucose to acetone, butanol
Lactobacillus sps.	Gelatin	Glucose to lactic acid
Solanum aviculare	Polyphenylene oxide -glutraldehyde	Production of steroids, glycoalkaloids
Streptomyces	Sephadex	Production of streptomycin

10 # HALOMETABOLITES

10.1 Introduction

Halometabolites are halogenated organic compounds, commonly found in nature and are produced by many different organisms. The halogen atoms are incorporated into a wide variety of organic compounds such as amino acids, indoles, hydrocarbons, phenols, terpenes, 15-C non-terpenes etc by enzyme catalyzed reactions with halide ions as the halogen source. About a millennium (1000) halometabolites are found in and have been reported from marine environment, produced especially by algae. In Table 10.1, the sources of halometabolites are tabulated.

Table 10.1 Halometabolites and their sources.

Bromometabolites

found in marine environment

produced by algae

examples include:

Bromocorodineol from *Sphaerococus coronopifolius*

(+)intricenyne from *Laurencia thyrsiflora*

Chlorometabolite

produced by terrestrial organisms

examples include:

withanolide D chlorohydrin from *Withania somnifera*

cistachlorin from *Cistanche salsce*

chloropterosin from *Pteridiun aquilinum*

7-chlorodentziol from *Mentzelia decaptula*

Iodo- and fluro-metabolites

produced infrequently

examples include:

fluorocitric acid from *Acacia georginae*

10.2 Biosynthesis of Halometabolites

The incorporation of halogen atoms into organic compounds by living organisms is an enzyme-catalyzed reaction. At least three different kinds of halometabolite producing enzymes have been discovered which include the following:

1. **Haloperoxidases**: The first halogenating enzyme was isolated from the fungus *Caldariomyces fumago* by L.P.Hager's research group in 1966. The enzyme requires hydrogen peroxide and a halide ion for halogenation of organic substrates suitable for electrophillic attack.

   The chloroperoxidase contains protoporphyrin IX as the prosthetic group. Similar haloperoxidases were isolated from mammals, marine invertebrates, algae and bacteria. In 1984, a novel type of haloperoxidase was detected by Vilter in the brown algae *Ascophyllum nodosum*. This bromoperoxidase is a non-haem haloperoxidase that requires vanadium for halogenating activity. Vanadium-dependent chloro- and bromo-peroxidases were isolated from lichen, algae and fungi. A different type of non-haem haloperoxidase was isolated from bacteria in 1988 that neither contains metal ion nor requires any co-factor for activity.

   The different types of haloperoxidases have one thing in common is that halogenation catalyzed by these enzymes lacks specificity. They show neither substrate specificity nor regioselectivity.

   The two schemes of classifying the haloperoxidase are presented in Tables 10.2 and 10.3.

Table 10.2 Classification of haloperoxidases depending upon constituent co-factor.

Haloperoxidase	Typical Example
Haem containing haloperoxidases	Horde raddish peroxidase (HRPO)
Non haem containing haloperoxidases requiring vanadium	Lactoperoxidase (LPO) algal bromoperoxidase (BPO) fungal chloroperoxidase (CPO)
Non haem containing haloperoxidases not requiring any co-factor	Bacterial bromoperoxidase (BPO)

Table 10.3 Classification of haloperoxidases based on the halide ion incorporated.

Common name	Source
Iodoperoxidase optimum pH=5 to 7	
Horse raddish peroxidase (HRPO)	horserasdish root brown algae (*Phaeoohyta*)
Thyroid peroxidase	thyroid gland of mammals and birds
Bromoperoxidases optimum pH = 5-7	
Bromoperoxidase (BPO)	red algae (*Rhodophyta, Bonnemaisonia lamifera*) green algae (*Chlorophyta, Penicillus capitatus*) brown algae (*Phaeophyta*) marine worms
	bacterium (*Streptomyces phaeochromogenes*)
Ovoperoxidase	sea urchin eggs
Lactoperoxidase (LPO)	secretions of mammals: milk, saliva, tears

Table 10.3 Contd...

Common name	Source
Chloroperoxidase optimum pH=2.5 to 3.5	
Chloroperoxidase (CPO)	fungus (*Caldariomyces fumago, Laminaria digitata, Phellinus pomaceus*)
	ice plant (*Mesembryanthemum crystallinum*)
Myeloperoxidase (MPO)	mammalian neutrophils
Eosinophil peroxidase (EPO)	mammalian eosinophil

2. **S-adenosyl methionine transferase (AdoMet):** This is a different type of enzyme involved in the formation of methyl halides. The involvement of AdoMet in halometabolite production seems to be restricted to the formation of single carbon metabolites.

3. **Halogenases**: Biosynthetic studies on the formation of halogenated antibiotics produced by bacteria and comparison of the structures of halometabolites show that halogenating enzymes involved in the formation of these metabolites must have substrate specificity and regioselectivity. Taking this into account, two halogenases in the bacterium *Ps. fluorescens* that belong to a novel class of halogenating enzyme were detected. The halogenases do not require hydrogen peroxide; instead they need $FADH_2$, produced from FAD and NADH by unspecific flavin reductases. In addition to $FADH_2$, oxygen and halide ions are necessary for the halogenation reaction. They are specific for their substrates and catalyze regiospecific halogenation of their substrates. Similar enzymes are also present in a number of other bacteria producing different halometabolites. Halogenases constitute the major source for halometabolite formation in bacteria and possibly in other organisms.

10.3 Mechanism

Though a large number of halometabolites have been reported and characterized, but the knowledge and research work elucidating the mechanism of their biosynthesis is scanty. So far some mechanisms that have been put forward include the following:

1. Haloperoxidase-mediated reaction in the presence of hydrogrn peroxide:

This is the known biological route for the formation of halometabolites. A general reaction is represented as follows:

$$\text{Carbon substrate} + X^- + H^+ + H_2O_2 \rightarrow \text{Halometabolite} + 2H_2O$$

Where the carbon substrate may be an alkyne, cyclopropane, phenol or aniline and X^- may be I^-, Br^-, Cl^- or a pseudohalide: thiocyanate SCN^-.

For example:

$$CH_3(CH_2)_4COCH_2COOH + Br^- + H_2O_2 \xrightarrow{\text{at pH} =5.8} CH_3(CH_2)_4COCH_2Br + CO_2$$

3-Oxooctanoic acid → 1-bromoheptanone

2. Haloperoxidase-mediated reaction through formation of hypophalous acid:

This is an alternative mechanism which suggests that a free hypophalous acid, such as HClO (hypochlorous acid) is the primary product of enzymatic catalysis and subsequently reacts with a broad range of substrates to yield a variety of secondary reaction products.

3. S-adenosyl-L-methionine (AdoMet)-dependent methyl chloride transferase (MCT)-mediated reaction:

The haloperoxidases catalyze the incorporation of halide ion into substrates having two or more carbon atoms, but not into one-carbon compounds like methane (CH_4) to produce CH_3X. An AdoMet dependent MCT has been detected that can methylate halide ion (X^-) to CH_3X. The enzyme catalyses the reaction in the following way:

$$CH_4 + X^- + AdoMet \xrightarrow{\quad MCT \quad} CH_3X + \text{S-adenosyl homocystein}$$
$$\text{AdoHcy}$$

So far only three halide methylating enzymes have been studied; one each from algae (*Endocladia muricata*), a fungus (*phyllobius pomaceus*) and a higher plant (*Mesembry anthemum crystallinum* and *Brassica oleracea*).

4. NADH-dependent halogenases mediated reaction:

The characteristic feature of halogenases is that they do not require H_2O_2 but need NADH. They are substrate specific and regioselective. They also require $FADH_2$ and O_2.

10.4 Significance

The halometabolites have numerous actions, which are categorized, in this section.

1. Defense/antimicrobial action: The halometabolites play a role in the defense mechanism of the organisms in which they are produced i.e. they may be used as antifeedants, insecticides, antifungal and antibacterial agents.

    The activity of a disinfectant increases on halogenation and that more the degree of halogenation, better the disinfectant activity. Examples of such halometabolites possessing antimicrobial activity include 2-chloro-1,6-dihydroxyphenazine from *Streptosporangium* sp.; 2-choloro- and 2,4-dichlorosulochrin from *Aspergillus terreus var. aureus*; laurinterol (a brominated metabolite) from marine algae *Laurencia pacifica*; prepacifenol from *Laurencia filiformis*; chodriol from *Laurencia yamada* and chloramphenicol.

2. '*Anti-*' activities: Apart from the *anti*microbial activity, halometabolites possess a wide variety of other '*anti-*' activities as shown in Table 10.4.

3. Biological activities: Some of the miscellaneous biological activities of halometabolites are presented in Table 10.5.

Table 10.4 '*Anti-*' Activities of halometabolites.

Anti-algal	Anti-fertilized sea urchin eggs
Anti-bindweed	Anti-grasshopper
Anti-black bean aphid	Anti-green algal
Anti-cancer	Anti-mustard
Anti-crab grass	Anti-mycotic
Anti-cyanobacterial	Anti-salt marsh caterpillar
Anti-feedant	Anti-spider mite

Table 10.5 Miscellaneous biological activities of halometabolites.

α- adrenergic activity	bronchodilation
β- adrenergic activity	muscle relaxation
cholinergic activity	hypothermic
phenobarbital potentiation	

4. Environmental significance-ozone depletion: Increasing atmospheric concentrations of chlorofluorocarbons (CFCs) and other halocarbons are thought to be the main cause of the appearance of the Antarctic ozone hole in the late 1970s and the ozone depletion observed over the Northern Hemisphere. However, natural halocarbons, particularly CHX_3, are also important halogen carriers in the atmosphere. This is especially true for $CHCl_3$, which despite its short atmospheric half life (1.2 years), is capable of reaching the stratosphere at concentrations high enough to make it the principal Cl^- carrier in the upper atmosphere. There, it competes with the major CFCs as a source of Cl atoms. $CHCl_3$ undergoes photolysis in the presence of OH^- radicals to form chlorine monoxide (ClO) and atomic Cl, each of which play a major role in the conversion of atmospheric O_3 to O_2 through the following reaction:

$$Cl + O_3 \rightarrow ClO + O_2$$
$$ClO + O \rightarrow Cl + O_2$$

The net reaction being:

$$O_3 + O \rightarrow 2O_2$$

Likewise, CH_3Br having an atmospheric half life of about 1.5 to 3 years, also converts O_3 to O_2. In the presence of both the halogens, O_3 destruction becomes even more efficient. The net reaction being:

$$2O_3 \rightarrow 3O_2$$

Despite abundant emission of CH_3I, its ability to contribute significantly to ozone depletion is considered minor because of its very short atmospheric half-life of five days.

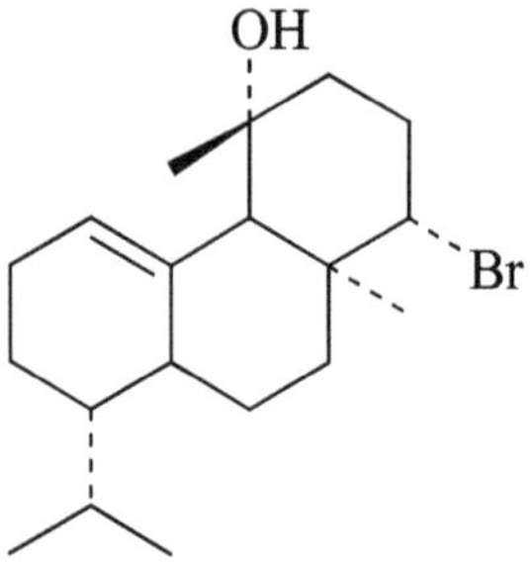

Bromocorodienol

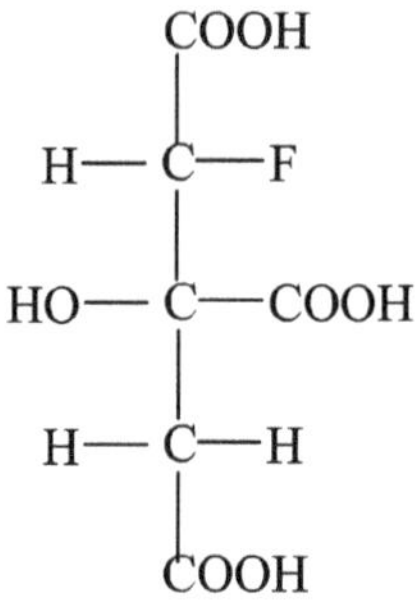

Flurocitric acid

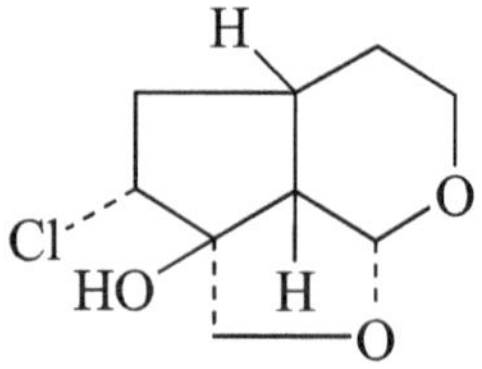

Cistachlorin

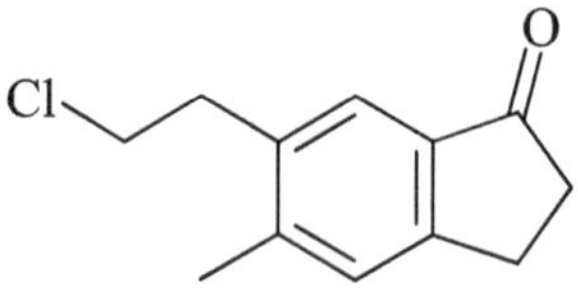

Chloropterosin

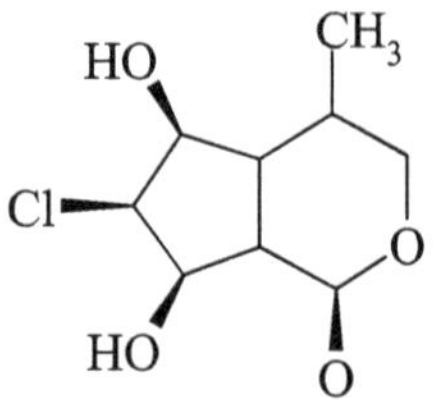

7-chlorodenziol

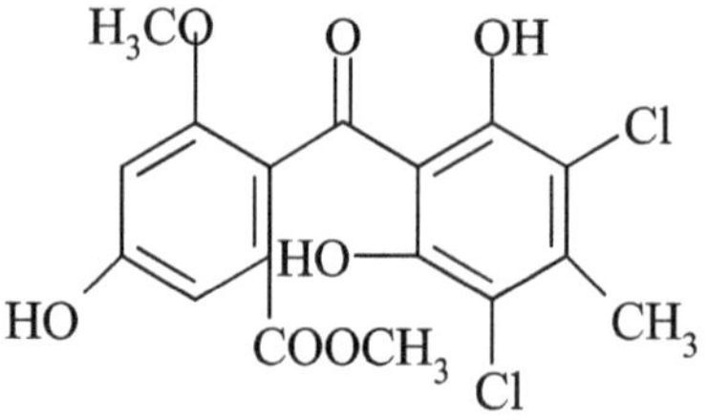

2,4-dichlorosucin

Figure 10.1 Representative structures of some halometabolites.

11   # BIOMASS: ENERGY PRODUCTION

CHAPTER CONTENTS AT A GLANCE

11.1 Introduction

Energy is a Greek word which means capacity to do work. It was first given by Thomas Young (1773-1829). The well-known, famous equation correlating mass with energy as given by Albert Einstein is:

$$E = mc^2$$

This relates to the fact that mass can be converted/ transformed into energy (and vice versa, is it possible!?!). That means you as a body, say in India, being transformed into energy and energy being transported at lightening speed, say to Japan, and retransformed back to mass i.e. you as a body in Japan!. However, what we are talking about and concerned with is the fact that at present the world population is heavily dependent on fossil fuels (natural gas, petrol, coal) for its energy needs. Alternatives to these so called conventional energy sources are being/have been worked out such as the use of nuclear (both fission and fusion), hydal, wind, solar energy etc. One such non-conventional energy source is bioenergy.

Bioenergy involves the use of biomass i.e. biological forms of matter such as plants, vegetables, their residues, animal excreta, microorganisms, enzymes etc that can be transformed into energy or can convert energy from one form to another.

Biomass may be categorized broadly into the following:

1. New plant growth: e.g. wood, short-rotation trees, herbaceous plants, conventional crops, algae, aquatic plants

2. Residues: e.g. crop/agricultural wastes, straws, husks, bagasse (the solid residue left after the juice has been squeezed out from cane sugar), corn cobs, bark, saw dust, wood shavings; cow dung, animal droppings

3. Wastes: garbage (solid trash that is thrown out), sewage (wash water that goes down the drain), sludge (the solid removed from the sewage treatment), industrial refuse such as sulphite liquor (from paper industry), slaughter house wastes, cotton seeds and fibres (from cotton industry), dairy industry wastes etc.

The biomass chemically consists of:

- polysaccharides e.g. cellulose, hemicelluloses, lignin, starch etc.

- water soluble constituents such as sugars, amino acids, peptides, aliphatic acids,

- ether and alcohol soluble constituents e.g. fats, oils, resin, pigments and

- proteins.

All these are transformable to energy and there is plenty of it available everywhere!

11.2 Bioenergy production: Methods

There are various ways, mechanisms, chemical reactions and techniques as to how the biomass can be transformed to bioenergy. Some of these are discussed in this section.

1. **Pyrolysis**: Pyrolysis is defined as the destructive distillation or decomposition of organic matter such as solid residues, wastes, saw dust, wood chips, wood pieces, cotton, bagasse, ground nut shell etc. in the absence of oxygen at a high temperature of about 200°C-500°C. Depending upon the process, heating rate and temperature, the products obtained on destructive distillation include gases, organic liquids and chars. The condensable liquids include pyroligenous acid, oil and tar; the gas being a mixture of carbon monoxide (28%-33%), methane (3%-18%) and hydrogen (1%-3%). Hydrogen content in the gas is directly proportional to the temperature, higher the temperature higher is the hydrogen content.

    Pyrolysis is mainly used to produce charcoal from wood used mainly for cooking purpose.

2. **Gasification**: It is a process of thermal degradation of carbonaceous material under controlled amount of air or oxygen and high temperature of up to 1000°C. The process reaction results in the formation of BTU gas consisting of carbon monoxide and hydrogen. The gas can itself be used as fuel or can be used as an intermediate for the synthesis of fuels or energy intensive chemicals such as methanol, substitute gasoline, hydrogen and ammonia.

3. **Bariquetting**: The process consists of drying cellulosic wastes using solar energy and then partially carbonizing the material to give ash-free carbon.

4. **Liquefaction**: In this process, wood or agricultural residues are reacted with carbon monoxide and water/steam at a high pressure of about 4,000 lb/in^2 and at a temperature of about 350°C-400°C in the presence of catalyst to produce oils for energy.

5. **Solar energy to bioenergy**: This involves the use of biological agency (its capacity to photosynthesize) to generate fuels or power either directly or indirectly from sunlight. In its simplest sense, this means growing plants which can be used as fuel (burning of wood).

    Alternatively, the electrochemistry of photosynthesis can be employed directly to generate electricity. This can be done either by using intact cells or by isolating photosystems (either I or II) which transduce light energy into electrochemical potential in the chloroplast or by using the antenna complex that captures the photons and passes on to a reactive centre.

6. **Biofuel**: It is a fuel generated from biomass such as starch, sugar, wood pulp, bagasse, sewage, waste etc. Examples of biofuel include gasoline, ethanol, methane, which can be produced by fermentation. These Biofuels are being used to run vehicles (using alcohol) or to generate electricity (using methane). Biodiesel, i.e. methyl ester of plant fatty acids, made by transesterification of plant oils (especially rapeseed oil) is also being used to run vehicles.

    Plants/ crops grown with the purpose of serving as raw material for generating fuel are referred to as energy plant/ crops.

7. **Anaerobic digestion- Biogas production**: Here, microbes convert organic matter (mostly sewage) by fermentation in the absence of oxygen into gases such as methane

and carbon dioxide which are collectively called as biogas. Fermentation is brought about by methanogenic bacteria in air-tight brick or concrete lined chambers called digestors. Methanogenic bacteria can utilize only simple substrates to produce biogas. Thus another population of bacteria that first breakdown the complex substrates are required. These are allowed to do their part of the job first followed by methanogenic bacteria. This is an example of consortium fermentation.

Cow dung is mainly used for the production of biogas. In first stage, it is degraded anaerobically by acid fermenting bacteria to free volatile short chain fatty acids such as acetate and propionate which subsequently in the second stage are converted by methanogenic bacteria to methane (approximately 60%) and carbon dioxide (approximately 40%). The digestor tanks may be up to 12,000 l capacity provided with heating coils for the temperature control. Some of the factors that favour the production of biogas are enlisted below:

- addition of algae (*zygogonium* sp.) to cow dung
- anaerobic conditions
- optimum pH of 6-8
- seeding with small amount of sludge from another digestor
- slurry concentration of 50% w/v and
- temperature: digestors can be operated either in mesophilic range of $25°C$-$38°C$ or in thermophilic range of $50°C$-$70°C$.

8. **Production of hydrogen**: Hydrogen, to be used as fuel, can be produced from almost any thing. Most commonly it is produced by the following ways:

- Reforming of methane or other hydrocarbons: The reaction involved is:

$$CH_4 + H_2O \xrightarrow{Ni^{2+}} CO + 3H_2$$

$$\downarrow{\scriptstyle Fe^+/Cr^{2+} \mid H_2O}$$

$$CO_2 + H_2$$

The liberated carbon dioxide is subsequently removed by physical; and chemical techniques.

- Synthesized from coal: The reaction involved is:

$$3C + O_2 + H_2O \rightarrow CO + H_2$$

- Partial oxidation of hydrocarbons: The reaction involved is:

$$CH_4 + 1/2O_2 \rightarrow CO + H_2$$
$$\downarrow$$
$$CO_2 + H_2$$

Carbon dioxide is subsequently removed.

- Biohydrogen- electrolysis of water using microbes: This is nothing but hydrogen, a biogaseous fuel, made by a biological system such as algae. The idea is to utilize organisms such as single-celled algae to breakdown water in the presence of light i.e. photolysis of water.

  Microbes such as *Scenedesmus, Chromatinum, Clortridium, Thiocapsa* and the recently discovered *Halobacterium* are the main groups of microbes possessing hydrogenase, the enzyme responsible for the generation of (bio)hydrogen.

  Enzymes in algae responsible for photolysis of water and release of hydrogen are oxygen sensitive i.e. the process does not occur in the presence of oxygen. Genetically engineered microalgae need to be created such as *Clamaydomonas reinhardtii* which are more oxygen tolerant.

  Nitrogenase enzyme of *Anabaena, Nostoc, Rivularia* etc. is an enzyme that reduces nitrogen into ammonia and releases hydrogen gas. Likewise, enzymes such as cellulose, glucose dehydrogenase and hydrogenase can be employed to release hydrogen from waste paper etc.

9. **Aerobic digestion-Composting**: It is defined as the biological decomposition of organic wastes involving their conversion by microbes in the presence of air to utilizable forms such as gaseous fuels, single cell protein, fertilizers etc.

10. **Hydrocarbons-Petroplants**: Plants of the family *Euphorbiaceae* especially and plants belonging to other families such as *Apocyanaceae, Ascelepidiaceae, Compositae, Dipterocarpaceae, Leguminosae, Moraceae* etc. have been screened and found to accumulate hydrocarbons of high molecular weight. These hydrocarbons closely resemble petroleum, crude oil or gasoline and therefore can be used as their substitutes. Examples of such petroplants include *Batryococus sp., Calotropis procera (aak), Chlorolla pyrenoidosa (algae), Copaifera sp. Dipterocarpus turbinatus, Hevea brasiliensis, Euphorbia sp., Parthenium arentatum*, etc.

The production and applications of bioenergy derived from biomass are presented in Table 11.1.

11.3 Conclusion

All in all, it can be said that a huge potential exists for converting the biomass into bioenergy involving the use of biotechnology such as genetically engineering the microbes, fungi, algae and plants. The production of biogas (methane), alcohol or the search for hydrocarbon accumulating petroplants is well established, but still R&D efforts are needed to develop novel techniques for bioenergy generation for screening potential organisms.

Just imagine one day, the *'pan spiting'* (a huge menace and waste material) could be collected in a small pencil-sized cell and a microbe in the cell could act on it to generate bioenergy which could be used to charge your cell phone! Or power your walk man! Could this become a reality someday?! No one knows, but that is biotech: innovative, novel, patentable, influencing your daily life!

Table 11.1 Bioenergy in action: The production and applications of bioenergy derived from biomass.

Country/Institute Company	Biological agency Biomass involved	Form of bioenergy	Applications
	Salvinia molesta (water fern) on aquatic biomass	biogas	-
	Eicchornia crassipes (water hyacinth) on aquatic biomass,	biogas	-
…heti Road, Sugar mill, …ijnor, U.P.	paddy, sewage sugarcane waste	-	generation of electricity
…hermal Power Plant, …lkheri, Punjab	paddy straw	-	"
…addy Processing …esearch Centre, Tiruvarur	husk	-	"
…entral Tuber Crop …esearch Institute, Trivandrum	starch from tuber of tapioca	alcohol	-
…ne third population of the world	fire wood	-	cooking, heating
…entral Mechanical Engineering & …esearch Institute, Duragapur	city garbage wastes,	biogas	generation of electricity
…dia	farm waste (manure)	BTU gas	used as fuel or as intermediate to produce other fuels such as methanol, gasoline hydrogen, ammonia

Table 11.1 **Contd….**

Country/Institute Company	Biological agency Biomass involved	Form of bioenergy	Applications
Apart from Brazil especially National Alcohol Program), America, Europe, Australia, Canada, Indonesia, Kenya	sugarcane	ethanol for energy called 'proalcool'	for running all Brazilian cars on ethanol-petrol mixture
US	corn starch	alcohol for energy called 'agricultural alcohol'	use of gasohol, i.e. 6-9 parts petrol and 1 part ethanol, for running cars
Bouche-du-Rhone, France	sugarbeet, Jerusalem artichoke (a genetically improved bioenergy crop)	agricultural alcohol	-
Japan	sugarcane	fuel alcohol	for substituting petrol
Biochemical Engineering Research Centre of IIT	cow dung	biogas	energy need of families, for replacing diesel
Okhla Sewage Disposal Works New Delhi	sewage	biogas	energy need of families
Gobar Gas Scheme, India	cow dung	"	"
Euphorbia Gasoline Refinery Italy	*Euphorbia lathyrisi& E. tirucalli*	hydrocarbons	as gasoline
Central Arid Zone Research Institute, Jodhpur	*aak*	"	as petroleum substitute
Novozymes	using cellulase on cellulose	-	-
France, Italy, Germany	esterification of plant oil	biodiesel	to run city bu

226

In the introductory chapter it was mentioned that biotechnology mainly encompasses the fields of fermentation which is referred to as 'traditional' biotechnology, plant culture, animal culture, genetic engineering and hybridoma technology which are referred to as 'modern' biotechnology. But in the 21st century, even the genetic engineering and hybridoma technology have in a sense become 'old'. At the onset of this century Human Genome Project (HGP) was completed (on 26th June, 2000). In the project, the entire human genetic code was deciphered. This genomic information is being used to cure, prevent or diagnose some 4000 genetic disorders in man. This is being made possible by employing 'novel' biotechnological techniques such as genomics, proteonomics, pharmacogenomics and bioinformatics. For instance, BRCA 1 (a gene the mutation of which is associated with a high incidence of breast and ovarian cancer in women) test, was the first diagnostic test product to be developed from the genome information.

In this 21st century, 'novel' biotech based techniques are already revolutionizing and improving the new drug discovery and development process. The 'novel' biotech based techniques include, apart from the ones as mentioned previously, genetically engineering the animals (transgenic, knock outs), protein engineering, peptide chemistry, peptidomimetics, nucleic acid technologies (antisense, aptamer, ribozyme catalysis), catalytic antibodies (abzymes), glycobiology and tissue engineering.

Rapid strides in one branch of science are always supported or paralleled by developments in other disciplines. For instance, rapid strides in biotech have been made possible due to fast changing developments in electronics and information technology industries. Developments of novel techniques such as DNA sequencing, cloning, polymerase chain reaction (PCR), protein extrusion, crystallography, 2-D (dimensional) gel electrophoresis, mass spectrometry, chip-based microarrays, protein NMR, phage display etc. have all contributed to the expansion of biological information. To better appreciate what has been stated, just consider this: the information of the entire genetic sequence generated by the HGP is so huge that if compiled in books, it would take 1000 volumes with each volume of 500 pages; further merely reading of the millennium volumes round the clock (twenty four x seven) it would take about 26 years. Alternatively, if the sequenced data is stored in computers, it will require a space of 10^9 bits i.e. 1000GB!!! (a PC has a space of 40GB; 25 such PCs would be required!) This is, as far as, genome sequence consisting of about 30,000 genes of one organism (human) is concerned. By now genomes of several hundreds of organisms have been sequenced ranging from five genes to over 100,000 genes. Actually, the data concerning biological macromolecules (such as proteins, genes) is increasing by leaps and bounds which has been made possible due to the use of high thorough put screening i.e. 'omics' like genomics and proteomics. The available data doubles in about a year and quarter! Now, how do we store this huge amount of data? Not only store, but organize it, make it accessible/retrievable world wide? Further, if one happens to find a piece of information about a biological macromolecule, then is it unique or someone has already worked this out i.e. one should be able to compare and analyze the stored data. Not only this, the information has to be translated (i.e. gene expression, determination of protein structure and their interactions etc.) and put to the welfare of mankind to develop new therapeutic and diagnostic agents and improve patient care. For handling all this biological information computers have become essential and a new disciple, bioinformatics, has emerged.

Bioinformatics is defined as the application of computational techniques to understand and organize the information associated with biological macromolecules. Stated literarily, it is a product resulting from the marriage of biotech with information technology or stated otherwise, it is *in silico* biology (biology on computer).

If one happens to be fascinated and interested in the field of bioinformatics, then one should have a sound knowledge of both the biological science (anatomy, histology, cell biology, biochemistry, genetics, molecular structure etc.) and computer science (programming, data structures, simple computer architecture, databases, computer networks, basic artificial intelligence, knowledge representation, qualitative modeling etc.) plus basic mathematics (calculus, differential equations, linear algebra, statistics etc.).

Bioinformatics therefore, comes to the rescue of a biological scientist who is flooded with lots and lots of information in the following ways:

1. Handling and management of biological data, including its organization, control, linkages, analysis etc.
2. Communication among people, projects and institutions engaged in biological research and applications
3. Organization, access, search and retrieval of biological information
4. Analysis and interpretation of the biological data through the computational approaches including visualization, modeling and simulation, development of algorithms etc.

The discipline of bioinformatics is based on the following:

1. **Genomics:** This is the comparative study of the complete genome sequence and its function from different organisms. There are three types of genomic studies:
    i.   **Structural genomics**: This involves determination of the nucleotide sequence of the entire genome of an organism.
    ii. **Functional genomics**: This involves determination of the function of each and every gene in an organism i.e. it involves determining the pattern of gene expression, mechanism by which gene expression is coordinated and interrelationship of gene expression when a change in cellular environment occurs.
    iii. **Comparative genomics**: This involves comparison of the structure, function, expression or product of genes of two different organisms. For instance, it has been found that gene expression and regulation of some genes in lower organisms resembles that in higher organisms like man. So the lower organisms can serve as true laboratory models concerning the particular genetic problem.

2. **Proteomics**: This involves determination of the structural, biochemical and physiological repertoire of all proteins.

3. **Data Retrieval System**: There are three major repositories world wide that publicly make available the bioinformatics data bases providing information concerning gene sequences, expression, regulation, protein structure etc. and tools for analysis. The three retrieval systems are:

i. **Entrez**: developed by National Centre for Biotech Information (NCBI), Maryland, US.

ii. **System Retrieval System (SRS)**: developed by European Biotech Institute (EBI), Cambridge, UK.

iii. **DBGET**: developed by Institute for Chemical Research, Kyoto University and Human Genome Centre of the University of Tokyo and is available through Genome Net.

In the Pharmaceutical Industry, bioinformatics has entirely changed the perspective of scientists engaged in discovery of new drugs and diagnostics and devising of individualized therapy (based on pharmacogenomics). Such a strategy based on the genomics is depicted in Figure 12.1.

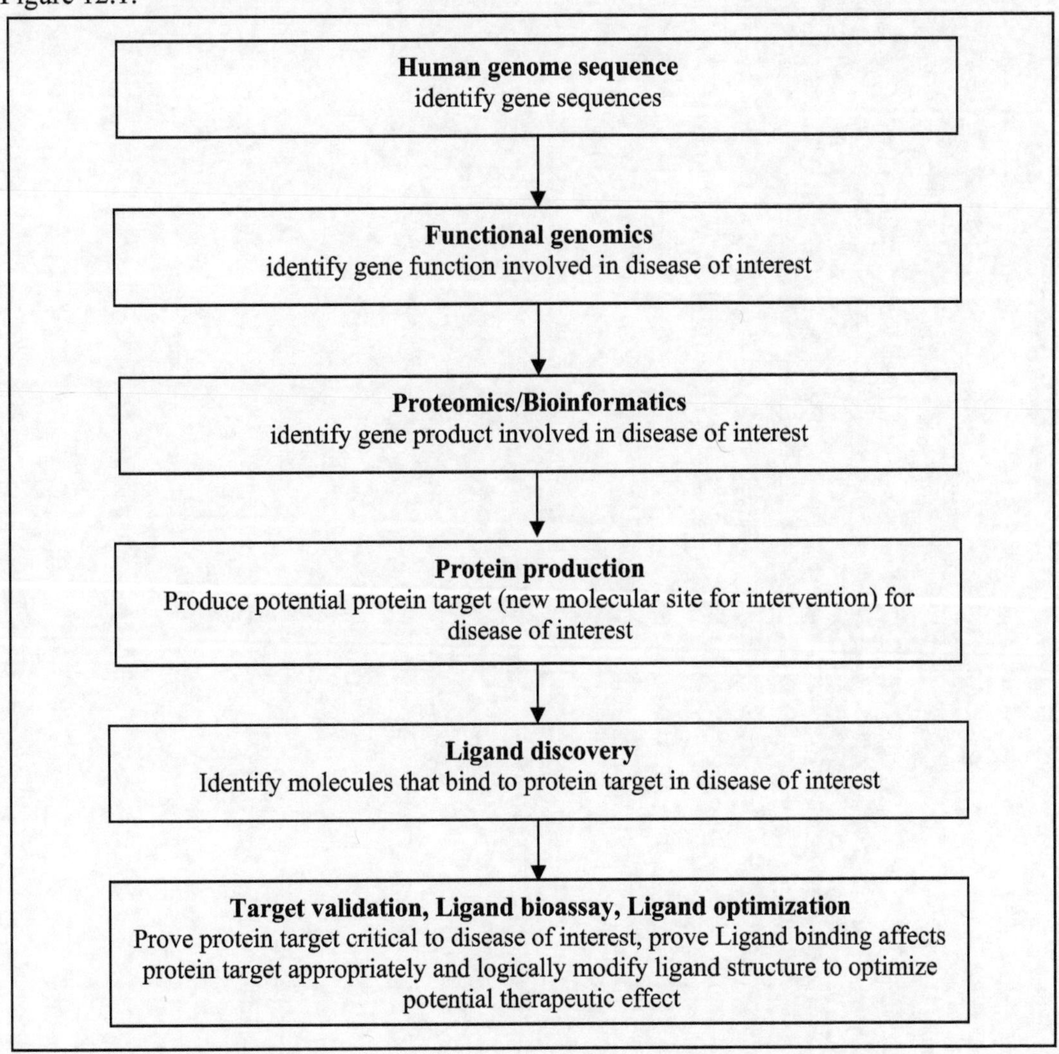

Figure. 12.1 New drug discovery and individualized therapy: Strategy of a bioinformatics!.

Large pharmaceutical houses have already set up R&D groups or 'virtual reality laboratories' working on the strategy of genomics/bioinformatics for molecular and drug designing. Completion of the HGP is only the beginning of an era in bioinformatics which is already revolutionizing drug developments, gene therapy and the entire approach to health care.

13 BIOTERMS, BIODEVICES AND BIOAPPLICATIONS

13.1 Introduction

The huge potential that the biological systems (viruses, bacteria, fungi, protozoa, plants and animals) and their parts have and which is being put to the benefit of mankind in one way or the other is being included here in the form of definitions of biodevices and bioapplications. Biodevices and bioapplications mean devices and applications which involve the use of biological systems. A collection of such terms is presented in Table 13.1. There are few devices and applications which need to be considered at length.

13.2 Biosensors

Biosensors are devices or probes that employ a biological element for measuring the concentration of desired substance.

13.2.1 Parts of biosensors

Principally biosensors consist of 2 parts:

1. ***Immobilized biological element***: It is the biological element that actually makes a contact with the substance being analyzed i.e. senses it and reacts with it and produces a weak signal of some sort that may be physical, chemical, electrochemical or optical. Biological elements include: immobilized enzymes, antibodies, antigens, DNA, cell organelles, microbial cells, plant or animal cells and even multicellular organisms such as Daphnia (a small fresh water shrimp).

2. ***Transducer/ detection and amplification system***: This is what that actually picks up i.e. detects, the weak signal generated by the biological element, amplifies it and converts the physical, chemical, electrochemical or optical signal into a detectable, electrical signal. The signal can be correlated with the concentration of substance being analyzed. The various detection systems/ transducers include: electrodes (electrochemical detection system), semiconductors, photon counters (thermal and mass detection systems, photometric detection, infrared, fluorescence and surface plasmin detection systems), sound detectors and piezoelectric devices. The principle of biosensors is schematically presented in Figure 13.1.

Table 13.1 Definitions and applications of some bioterms.

Term	Refers to (-involving use of biological system/agency)	Application/ Remarks
Bioaccumulation Related terms (RT): Bioremediation, Microbial mining Biosorption	accumulation of metals from surroundings	mining of metals present in very low concentration, removal of toxic metals from water using microbes
Bioassay	determination of concentration of a chemical	BOD test, quantification and identification of cytokines, growth factors present in minute quantities
Biocatalysis RT: Biotransformation Bioconversion	conversion of chemicals	numerous reactions and products such as steroids, prostaglandins etc.
Biochip RT: Gene chip, Microarray, DNA array	a computer chip that uses biological element as part of its construction made using micro engineered mechanical systems	construction and functioning of microdevices i.e. making laboratory ware and laboratory analysis a reality at microscale; commonly called 'lab on chip'
Biocontrol RT: Biopesticide	a planned introduction of one species that outgrows another	to control destructive insects
Biocorrosion	corrosion brought about by microbes	detrimental!
Biocosmetics RT: Cosmeceuticals	cosmetics having biological ingredients such as collagen, hyaluronic acid, gamma linoleic acid, retinoid etc.	protectants, overcome wrinkles and so on and so forth
Biodegradable materials	broken down by microbes (soil bacterium)	pollution control

Table 13.1 *Contd…*

Term	Refers to (-involving use of biological system/agency)	Application/ Remarks
Biodiversity	diversity of life forms	bioprospecting
Biodrug	drug produced by microbe, in human body at the site of action	delivery system with enhanced stability and potential for drug targeting
Bioethical RT: Ethical, Legal and Social Issues (ELSI)	the 'good factor' or morale regarding use of biological systems	so that man remains within his limits (reticular formation!)
Biofilm RT: Biofouling, Biofilter, Biosensor, Bioimmobilization	a layer of microbes growing and/or fixed on a surface	all the related terms are bioapplications
Biofilter	microbes supported on the surface of a filter equipment	used for waste disposal/ treatment, wherein microbes degrade the chemicals/ sewage in fluid flowing through the film
Biofuels RT: Biocrop, Biogas	use of biological material or systems to generate fuel	energy production from biomass
Biohydrometallurgy	use of bacteria to perform processes involving metals	microbial mining, oil recovery, desulphurization, redox reactions
Bioinformatics	collection, storage, organization, correction (i.e. updating), analyses of information concerning biological systems	drug discovery, diagnosis of disease and its pathogenesis
Biolistics RT: Gene gun	particle gun	shoots (introduces) DNA into any living cell

Table 13.1 *Contd.....*

Term	Refers to (-involving use of biological system/agency)	Application/ Remarks
Biological response modifiers (BRM) RT: cytokine	protein that affects immune functioning	e.g. interleukins as anticancer agents
Bioenhancer	an agent that enhances the absorption of drugs	dose of drug is reduced (e.g. cow urine!)
Biological warfare agricultural RT: Bioweapons, Bioterrorism, Bioarms Biological bomb	deliberate engineering of super potent infectious agents to be used as weapons	to attack an enemy's population (e.g. *Bacillus anthrax*), plants or domestic animals
Bioluminescence	production of light from biological systems	as (bio)-detection system/ biosensor; having ornamental/ aesthetic value
Biomass RT: Single Cell Protein	bulk biological mass	as food (spirullina); energy production
Biomimetics	synthetic chemical agents able to perform biological functions	fascinating applications!
Biopharmaceuticals RT: Biologics, Pharmaceutical proteins	Pharmaceuticals of biological origin (proteins); Companies using biotechnology and advanced research tools (genomics) to discover/ make new and better molecules; fate of drug inside the body	amongst number one field in Pharmacy
Biopolymers RT: Bioplastics, Biopol	polymers made of biomaterial or made by biological systems	artificial tissues, wound dressings (of hyaluronate polymer), dextran for chromatography, polyhydroxy alkonates as plastics

Table 13.1 *Contd...*

Term	Refers to (-involving use of biological system/agency)	Application/ Remarks
Biopreservation	preservation of food using biological materials i.e. a bacterium killing another one	food preservation
Bioprospecting	search for new biological resources from which natural products could be derived	for commercial production of natural products
Biopiracy	'stealing' away biological material from a country	detrimental to native country; beneficial to the other!
Bioreactor RT: Fermentor	a vessel in which biological reaction or change takes place	fermentation, biotransformation
Biosensor RT: Bioelectrodes	a device that determines analyte concentration based on biological element	quantification of minute quantities
Biotransformation	conversion of one chemical into another brought about by a biological agent	production of steroids, etc.
Biotechnology	use of biological systems for fermentation, genetics engineering, tissue culture, biotransformation and diagnosis	enhancement of biological capabilities; numerous products and applications
Biosurfactant	a surfactant produced by microbes	used for their surface active property in various industries, oil recovery, foods, cosmetics and pharmaceuticals
Biotin	Is it some form of tin produced by some microbe? Oh! No, it is a Vitamin!!!	as cofactor

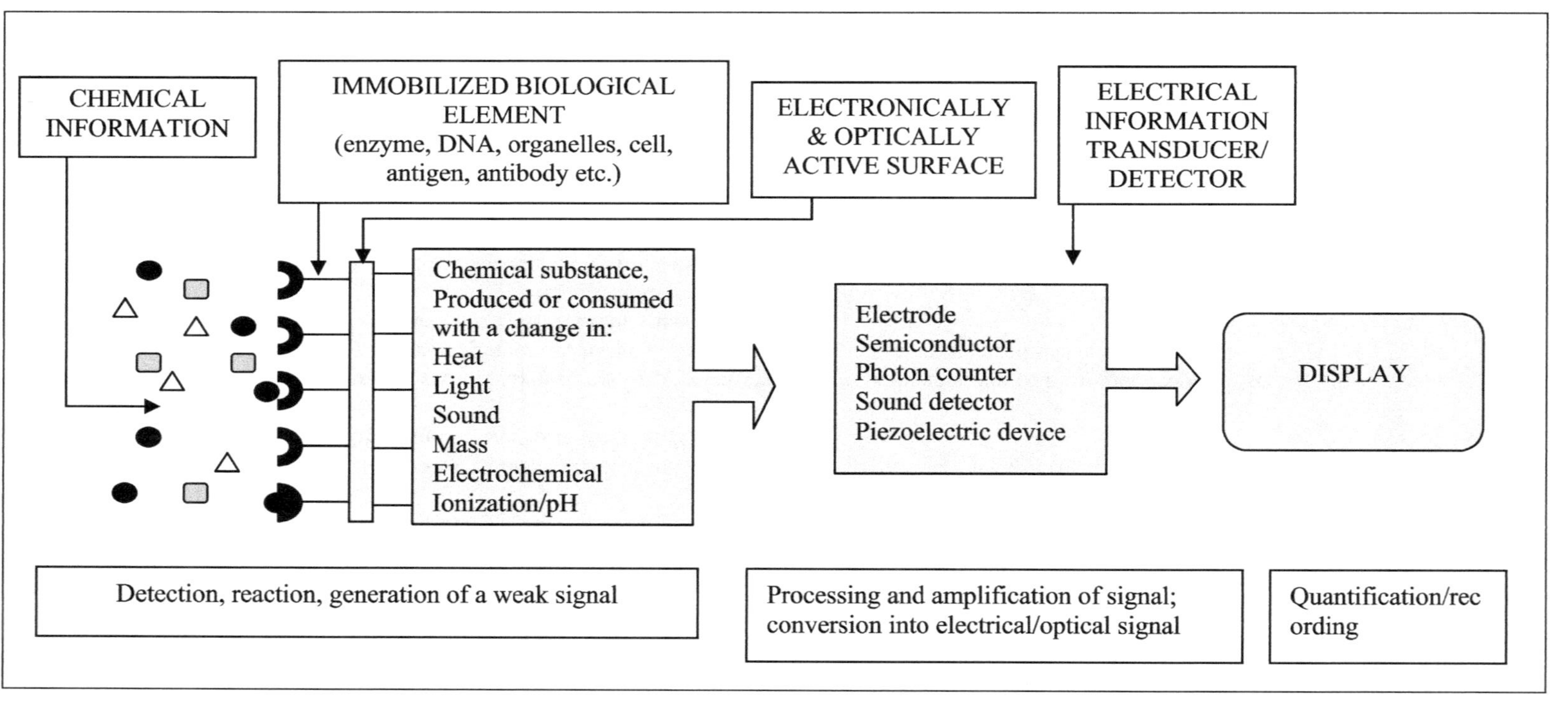

Figure 13.1 Principle of biosensors.

13.2.2 Kinds of Biosensors

Based on the biological element or the detector system used, biosensors can be of different kinds. The different kinds of biosensors include:

1. *Enzyme electrodes*: In this type, an enzyme is immobilized onto the surface of an electrode. The enzyme is immobilized by one of the following ways: physical adsorption, chemical cross linking, encapsulation in a gel or entrapment in polymeric membrane.

When the immobilized enzyme on the surface of the electrode is dipped into the analyte/ substrate solution, the enzyme does its job i.e. it catalyzes the formation/ consumption of product/ substrate accompanied with generation of electrons. These electrons are transferred on to the electrode surface, resulting in generation of current.

The enzyme electrodes are of 2 types:

    i. *Amperometric:* i.e. those which measure the current as a result of flow of electrons

    ii. *Potentiometric:* i.e. those which measure the voltage generated by the reaction. It is the voltage that is to be applied to counter balance the generated voltage.

Practically, speaking only one enzyme electrode i.e. glucose biosensor, for monitoring the blood glucose concentration in diabetics, is a commercial success.

2. *Electrochemical biosensor*: Here the biological element (most commonly an enzyme) generates a distinct chemical product such as an acid, which is then detected by the electrode.

3. *Ion-Sensitive-Field-Effect Transistor (ISFET)*: These are fabricated using semiconductor technology, using which ISFTE can be miniaturized and integrated on one chip. Hence, they are also known as microbiosensors. A semi-conducting device is one in which the electric field over n-p-n or p-n-p junction modulates the current flowing through that junction. The electric field over the junction is created by a build up of ions near it. As a result of this current flows through the 'gate' and the current flowing is dependent on the ions present (see Figure 13.2). The device is, hence, named Ion-Sensitive.

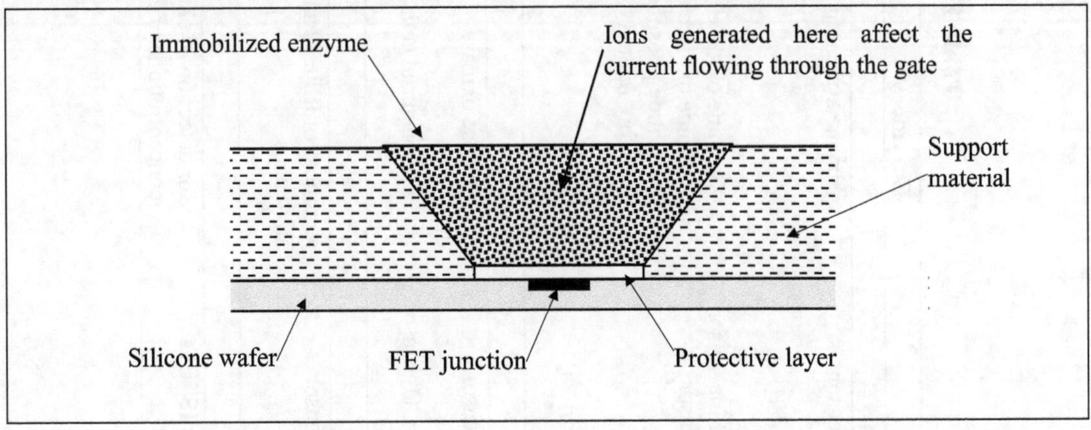

Figure. 13.2 Schematic presentation of Field Effect Transistor.

Table 13.2 Examples of some commonly used EnzFET.

Sensor	Enzyme	Application
		pH electrode
Alcohol sensor	enzyme system in cell membrane of	acetic acid- converts alcohol into acetic acid via acetaldehyde producing bacterium: includes alcohol dehydrogenase, aldehyde dehydrogenase, electron transferase
Hypoxanthine sensor Inosine sensor	Xanthine oxidase Xanthine oxidase co-immobilized with nucleotide phosphorylase therefore related to the freshness of fish.	converts hypoxanthine/ xanthine/ inosine, decomposition products of fresh fish (ATP→ ADP→ AMP→ IMP→ inosine→ xanthine), to uric acid. Measure of hypoxanthine levels is
Urea sensor	Urease	converts urea into uric acid (pH change)
		Micro-Oxygen electrode
Glucose sensor	glucose oxidase	determination blood glucose concentration
Microbial CO_2 sensor	S17 bacterial cells	-
		Integrated multibiosensors
Multibiosensor	a number of ISFETs and microelectrodes are integrated into a biosensor capable of analyzing a number of substrates all at ones	
		Miscellaneous
Numerous ISFETs	for detection of chiral amine salts, choline esterase enzyme inhibitors, anionic surfactants; for determining concentrations of anesthetics: procaine, lidocaine; for monitoring intramyocardial pH	

Now, if enzymes (or proteins or cells) that generate ions on reacting with the substrate are immobilized near the junction, then there shall be a build up of ions near the junction. This would result in flow of current that can be related to the concentration of substrate. Such a device is called Enzyme-Field-Effect-Transistor (EnzFET or ENFET). Since ISFET is used as potentiometric transducer, enzyme such as urease and oxidase, which cause pH changes in their reaction are commonly used to make EnzFET. Presented in Table 13.2 are examples of some commonly used EnzFET.

4. *Immunosensors:* In this type of biosensor, the biological element is the antibody, that detects and reacts (binds) with the analyte. The binding triggers a reaction that is to be detected and quantified to give a measurement of the concentration of the analyte. The detection systems are one of the two kinds i.e. either mass-based sensors or optical devices. Mass detectors are manufactured on silicone chips (hence also called microchip biosensors). These detect the very small change in mass that takes place on binding of the analyte with the antibody. These are mass detectors that are based on surface acoustic waves (SAW). The tuning fork in these devices is made of piezoelectric material; hence they are also called as piezoelectric sensors.

5. *Optical biosensors:* In these type of biosensors, the biological element (most commonly or rather the only one employed) is the antibody. The effect of binding between the analyte and antibody is detected using light i.e. optical based detection system. The optical sensors are based on one of the two principles: i.e. either absorption of evanescent waves (i.e. transient, short-lived leakage of light across the fibre edge) or surface plasmin resonance (SPR) (creation of energy waves or 'plasmons' when light is scattered off a conducting surface at the correct angle).

The advantage of an optical sensor (fibre) is that it is very small in size and has a considerable strength. It can be inserted into a vein for detection of certain chemicals (i.e. Fibre Optical Chemical Sensor-FOCS). Well known examples of FOCS include: pH, oxygen and carbon dioxide sensors.

6. *Thermal sensors:* These are categorized as physical sensors along with mass sensors. In these type of biosensors, the biological element is the enzyme. When an enzymatic reaction occurs, it is accompanied with the release of heat. (This is more often so in almost all the reactions than the transfer of electrons required for electro-chemical detection!). But of course, any success or commercialization of thermal sensors requires the need of having very-very sensitive heat sensors as the heat evolved in a dilute solution is very meager.

7. *Patch clamp or Ion-channel sensors:* Ion-channels are proteins present in biological membranes (which are hydrophobic, lipoidal in nature) that allow the passage of ions (which are hydrophilic, aqueous in nature) across the membranes as a mechanism of regulating physiological functions.

Thus, in a patch clamp sensor, the biological element is the bilayer membrane having the ion channel and the electrode detects and measures, electrically, the flow of ions across the membrane through the ion channel. An example of the noted protein serving the purpose of forming ion channels in biological membranes is α-hemolysin.

Also, the ion channels (proteins) or the cells containing them can be immobilized onto an ISFET!

8. *Immobilized cell biosensors:* In these type of biosensors, the biological element is a cell; be it bacterial, fungal, algal, plant or animal cell; though, amongst the most common are bacterial cell biosensors. The reaction between the cell and analyte results in generation of signal that is detected by a detector system which may be based on diverse mechanisms. The diverse detection systems in cell (bacterial) biosensors include:

- Measurement of gas (O_2, CO_2) generation or depletion, using oxygen or carbon dioxide electrode or pH electrode.

- Measurement of luminescent light produced by luminescent bacteria or GM (genetically modified) luminescent bacteria on reacting with the analyte

- Measurement of flow of current by joining the electrode directly into the bacterial electron transport system i.e. direct electro-chemical coupling.

13.3 DNA fingerprinting

DNA fingerprinting is a technique for identifying human individuals based on a restriction enzyme (RE) digest of tandem repeated DNA sequences that are scattered throughout the human genome but are unique to each individual. In criminal cases, finger prints of the culprit are used to establish the identity of the culprit; just the same way, DNA 'print' is used for the same purpose. The DNA fragment is treated with sequence specific RE and the resulting short segments are sequenced base by base. The sequence is then compared with the sequence of DNA sample collected from the individual whose identity has to be established. Except for identical twins, the DNA 'print' is unique for each individual.

13.4 Polymerase Chain Reaction (PCR)

PCR is a novel procedure, using which it has become possible to produce multiple (millions and billions of) copies *invitro* (in Ependroff's tubes) using a very scanty amount of DNA. This is one of the most commonly known and widely used procedures categorized amongst the DNA amplification methods. Apart from PCR, other DNA amplification methods include: Ligase Chain Reaction (LCR) and Nucleic Acid Sequence-Based Amplification (NASBA).

PCR was invented in the year 1989, by Kary Mullis, working in Cetus Corporation, a biotech based US Company. Amplification of DNA, based on PCR is depicted schematically in Figure 13.3.

Stepwise procedure involves:

1. ***Denaturation (High temperature):*** The strands of genomic DNA to be amplified are separated by heating to a temperature of $90^{\circ}C$-$98^{\circ}C$.

2. ***Hybridization (Low temperature):*** On lowering the temperature to $40^{\circ}C$-$60^{\circ}C$, two short DNA molecules which serve as primers, bind to their complementary strands on two sites either side of the piece of genomic DNA to be amplified.

3. ***Replication:*** In the presence of a DNA polymerase and provision of all four nucleoside triphosphates (dATP, dCTP, dGTP, dTTP), synthesis of complementary strands takes place.

These three steps constitute one cycle. Thus, at the end of first cycle, number of copies generated is $2^1 = 2$; at the end of cycle number 2, number of copies is $2^2 = 4$; likewise for 3 cycles, $2^3 = 8$; 4cycles, $2^4 = 16$; and so on and so forth. One cycle (with time fixed for each step) ranges from 1-3 minutes. Considering 30 cycles and cycle time of one minute, one can have $2^{30} = 1,073,741,824$ copies of DNA in just about half an hour!!!

The DNA polymerase used in PCR needs to be thermostable. Such thermostable DNA polymerases include:

i. Taq DNA polymerase isolated from *Thermus aquaticus*

ii. Pflu DNA polymerase isolated from *Pyrococus furiosus*

iii. Vent DNA polymerase isolated from *Thermococus litoralis*

The piece of equipment that is required to carry out the PC is called thermal cycler. It automatically performs the stepwise procedure, varying the temperature from high to low in each cycle over 20 cycles – 60 cycles.

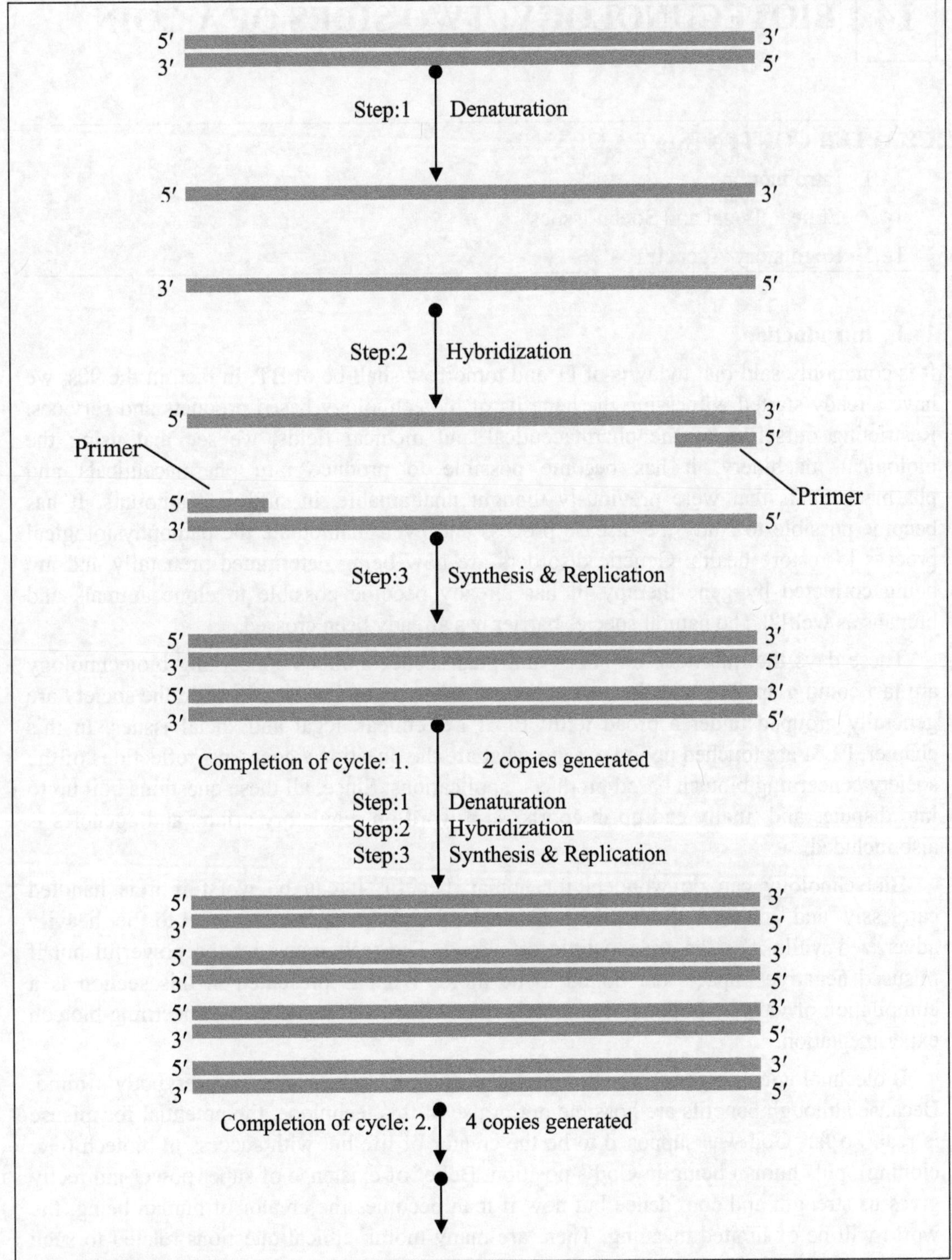

Figure. 13.3: Amplification of DNA, based on Polymerase Chain Reaction (PCR).

14

BIOTECHNOLOGY: TWO SIDES OF A COIN

14.1 Introduction

It is commonly said that today is of IT and tomorrow shall be of BT. In fact, in the 90s, we have already started witnessing the benefits of biotechnology based products and services. Restricting ourselves to the pharmaceutical and medical fields, we see that using the biological machinery, it has become possible to produce rare pharmaceuticals and pharmaceuticals that were previously thought unattainable, in sufficient amounts. It has become possible to study the disease process and even manipulate the pathophysiological process to restore health. Genetic disorders are now being determined prenatally and are being corrected by gene therapy. It has already become possible to clone animals and humans as well!?! The natural species barrier has already been crossed.

These days the questions on ethics and pharmacoeconomics concerning biotechnology are fast coming up. The questions on ethics i.e. what is good or bad for or in the society are generally grouped under a broad term: ELSI i.e. ethical, legal and social issues. In this chapter, ELSI are touched upon i.e. as to what are the questions/ concerns/ reflections of the society concerning biotech based products/ applications. Since, all these questions boil up to into disputes and finally end up in courts, so a word on regulatory affairs and agencies is also included.

Biotechnology can do wonders for human benefit, it can be worst if it is handled carelessly and misused deliberately. Biotechnology's benefits continue to be heavily advertized while its risks are too little discussed. The techniques are too powerful but if misused negative impacts are bound to be there. What is presented in this section is a compilation of views/ opinions of experts/ organizations the world over concerning biotech experimentation.

Biotechnological experiments have created an anxiety and fear in everybody's mind. Because although benefits are possible out of use of this technique, the potential for misuse is real. So far, God was supposed to be the creator of life but with success in biotech (e.g. cloning), puts human being in God's position. Belief of existence of super power indirectly gives us strength and confidence but now if man becomes the creator of human being, the world will be of limited meaning. There are many moral, ethical questions related to such experiments and their success.

Quoted below are the views concerning biotech research:

"Biotechnology promises the greatest revolution in human history. By the end of this decade, it will have outdistanced atomic power and computers in its effect on our everyday lives. But the biotechnology revolution is uncontrolled. No one supervises it. There is no coherent government policy, in America or anywhere else in the world. No federal laws regulate it. Genetics research continues at a more furious pace than ever. But it is done in secret and in haste and for profit."

- Michael Crichton

Author of Jurassic park in introduction to the book.

"The hub of the new (genetic engineering) technology is to move genes back and forth, not only across species lines, but across any boundaries that divide living organisms – the results will be essentially new organisms, self perpetuating and hence permanent. Once created, they cannot be recalled."

- George Wald, Nobel Laureate

"The very power of the new technology (biotechnology) outstrips our capacity to use it in safety, that neither nature's resilience nor our own institutions are adequate protection against the unanticipated impacts of genetic engineering."

- Licbe Cavalier, George Wald & David Suzuki

Prominent Scientists.

14.2   Ethical, Legal and Social Issues

Thus at various levels, ethical, legal and social issues concerning the use of biotech are posed. Some of these issues are considered in this section. However, these are opinions/ fears/ cautions of experts/ critics/ NGOs round the world. But what shall be the outcome of using biotech based products and GMO (genetically modified organisms) can only be concluded by carrying out intensive experimentation. It is only time that will tell us whether we have opted for the right or wrong (ethics!).

14.2.1   Ethical issues

The use of biotechnology creates profound ethical questions.

- Should we become architects of life itself? Should biotechnology be allowed to play god? Whether crossing the species boundaries in genetic exchanges, i.e. inserting animal genes into human or human genes into animals and inserting plant genes into microorganisms and other species is correct?

- Patenting genetically engineered animal is equating it to a state of manufactured product! Will living things have no more intrinsic value than automobiles or garments or any other commodity?

- The failure to label genetically engineered foods means that persons who follow religious dietary restrictions will be unable to ensure compliance with their beliefs. The genes from prohibited foods could be engineered into other foods. Thus genetic material whose consumption violates religious restrictions may be present in vegetable, fruit, etc. Consumer will not know which processed foods are genetically engineered. For example, vegetarians may be forced inadvertently to take food containing genetic material from insects, fish, pigs or other animal. Religious and ethical beliefs will be greatly disturbed.

- Would mothers be willing to buy infant formula with bioengineered ingredients extracted from udders of transgenic cow? (Human lactoferrin and human lysozyme are expressed in cow's milk.)

- Should humans be genetically engineered? Gene therapy should be restricted to the alleviation of genetic diseases in individual patients and should not be used to change or enhance normal human traits. Genetic modification of reproductive cells or the germ cells should not at present be attempted.

- Attempt to patent human DNA sequence which is our common heritage is an ugly side of research on human genome.

14.2.2 Legal issues

The most serious aspect of biotechnology is perhaps lack of supervision, lack of regulation, secrecy, patenting, testing the production in countries which are ignorant of effects, illegal release of genetically engineered organisms and many such issues which need more attention. Some such issues are presented below.

- Quest for patents may inhibit exchange of ideas amongst even research scientists.

- Transfer of most of the research in biotechnology from the public sector to private companies erodes the public accountability so necessary in the research and development. There will be least concern of public interests.

- A new generation of high tech foods, the product of genetic engineering technologies, which has reached consumer, has no legal binding from the US Food and Drug Administration of labeling (genetically engineered) and safety testing.

- Genetic engineering is being used to revolutionize biological warfare through creation of novel viruses and bacteria, which could have catastrophic effects and initiate a genetic arms race. Even AIDS virus is supposed to be outcome of such research. There has to be legal ban and appropriate control to stop such applications.

14.2.3   Social issues

- Will benefits of molecular biotechnology be available to wealthy only or all? The doubt is obvious while looking at the investments and expenses done on various research projects. Health, food, disease prevention and the problems of third world countries get less priority. More money is spent on combating rare genetic disorders than on research for malaria vaccine; or for instance potato is engineered to produce plastic from starch while potato as food is not available in some poor countries.

- Bovine growth hormone (BGH) injection into cows can increase milk production by 30% and prices may fall by 10%-15%. 25%-30% reduction in dairy farmers will be required to restore the market equilibrium. Also shift in milk production from smaller dairy farms to larger ones will be the consequence.

- Finance to molecular biotechnology will cause constraint on other important technologies.

- BT cotton

14.2.4   Environmental issues

This is one issue concerning which effects of molecular biotechnology work are not proved but there are many doubts pertaining to environmental safety. The pertinent questions include:

- Are genetically engineered organisms harmful to other organisms or environment?

- Introduction of novel genes into foreign habitats may disturb the natural equilibrium. Carp, catfish, trout and salmon have been engineered with a number of genes from humans, cattle and rats to increase their growth and reproduction. If released into environment the novel mutant fish could mate with native species, polluting the gene pool of native species quickly and permanently.

- Genetic engineering of organisms and products of organisms on large-scale with certain characteristics will reduce natural genetic diversity. This will result into genetic uniformity in farm animals and in agricultural plants. Prolonged use of cross genetic species may result into loss of natural world.

- Release of genetically manipulated organisms (GMOs) pose a risk because microorganisms in lakes, rivers, seas are capable of swapping much more genetic material than expected. Also phages are feared to play a role in transfer of genetic material from GMOs.

14.2.5   Health and Safety Issues

Genetic engineering has proved to be unsafe in some examples and some projects have been even abandoned due to negative results. Further word on this would be unjust/ inappropriate since it is too early to jump to conclusions on this issue.

What can be said is that biotech is new and is advancing rapidly; therefore, evaluation of rewards and risks is difficult. Resultantly, there are some who have hailed gene technology as timely contribution others have added a note of caution and have preferred to be optimistic. No one wants to wait for 'genetic Chernobyl' or 'genetic Bhopal' to tragically occur for realization of inherent dangers. The technologies are not to be blamed but it is the misuse which has to be controlled. Risk assessment and licensing of the experiments involving introduction of DNA in organisms exists. Some of the authorities/ governmental committees concerned with biotech issues are presented in the next section.

14.3 Regulatory Agencies

Rules concerning the manufacture, import, storage, processing, packaging and re-packaging of microbial and gene-technology based products and the research work involving microorganisms, genetically engineered organisms or cells have been framed by the Ministry of Environment and Forests, Government of India. A handful of governmental committees, functioning in the Department of Biotechnology, exist for monitoring the biotechnology based products and R&D work. Some of the noted committees include:

1. **Recombinant DNA Advisory Committee (RDAC):** The job of this committee is to review the developments in biotechnology at national and international levels. In India, the research work in recombinant DNA technology including its use and applications are governed by the regulatory recommendations of this committee.

2. **Review Committee on Genetic Manipulation (RCGM):** The look out of this committee is the safety aspect of the ongoing research projects and activities involving the use of GM (genetically modified) microbes.

3. **Genetic Engineering Approval Committee (GEAC):** This committee is concerned with the impact of releasing microbes in the environment. Approval of industrial or research projects involving the use of hazardous/ GM microbes at a large scale and having environmental concern, comes under the purview of this committee.

4. **State Biotechnology Coordination Committee (SBCC):** This state-level committee has the powers to inspect and investigate the safety and control measures adopted by the organizations/ institutions taking up projects/ research work involving the use of hazardous/ GM microbes and take action against them if found non-compliant with the norms.

5. **District Level Committee (DLC):** This is a committee, functioning under the Collector at the district level, which submits its report of inspection of the biotech installations in the district to the SBCC or GEAC for further approval and to meet any emergencies.

6. **Institutional Biosafety Committee (IBSC):** This is a committee that is constituted within an academic institution engaged in handling of microbes/ hazardous microbes/ GM microbes (just like the institutional animal ethics committee).

Figure 14.1 presents the biohazard symbol which can be seen on the laboratories handling biohazardous materials with the instruction that admission is restricted to authorized personnel only.

Figure. 14.1: The Biohazard symbol.

QUESTION BANK

CHAPTER 2: GENETICS AND GENETIC ENGINEERING

ESSAY TYPE QUESTIONS

1. (a) Explain recombinant DNA techniques.
   (b) Describe determination of sequences of bases in DNA.

2. Discuss in detail replication of DNA and special characters of DNA.

   What is recombinant DNA technology? Discuss the use of recombinant technology in drug therapy.

4. Write short notes on:
   (a) Genetic engineering
   (b) Molecular properties of DNA
   (c) DNA as genetic material

   Explain the recombinant DNA techniques. Describe various steps involved in DNA replication.

6. What do you understand by replication of DNA? Describe various base pairs and their properties. Describe the role of base pairs in semi-conservative process of DNA replication.

7. (a) Describe the recombinant DNA technique.
   (b) Discuss the production of interferons by recombinant DNA technology.

8. Write short notes on
   (a) Strategies of foreign gene expression
   (b) Explain-DNA as a genetic material

9. Write short notes on
   (a) Production of growth hormone
   (b) Replication of DNA

10. Write an exhaustive note on DNA polymerases.

249

11. Write a brief note on the following:
    (a) Use of plasmid as vector   (b) Use of cosmid as vector
    (c) cDNA   (d) Restriction endonucleases
    (e) Prokaryote *vs* eukaryote as expression host cell
    (f) Humulin®   (g) Activase®
    (h) Streptokinase

OBJECTIVE TYPE QUESTIONS

A. Multiple choice questions

1. The most important clue that helped in the determination of double helical structure of DNA came from
    (A) Chargaff's rules
    (B) Hershey-Chase experiments
    (C) Avery-MacLeod-McCarty experiments
    (D) Nirenberg and Khorana codon assignment
    (GATE 2005 Pharmaceutical Sciences)

2. Polynucleotide kinase is used
    (A) to add a nitrogenous base at the 5′ end of DNA
    (B) to add a nitrogenous base at the 3′ end of DNA
    (C) to add a phosphate at the 5′ end of DNA
    (D) to add a phosphate at the 3′ end of DNA   *(GATE 2006 Life Sciences)*

3. The organization of an eukaryotic gene expressed at high levels in liver is diagrammatically represented below:

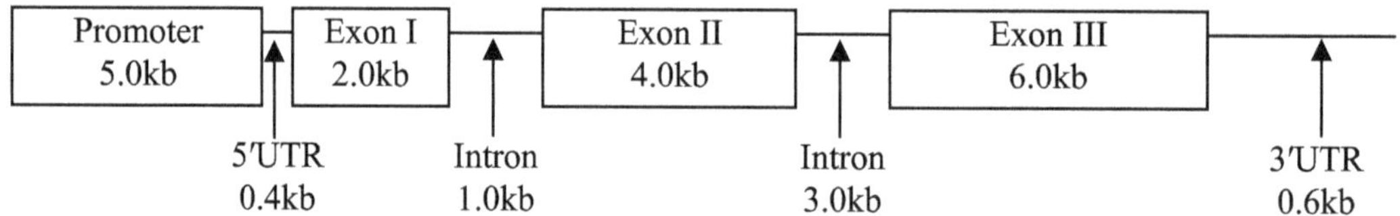

    The size of the mature mRNA generated by the transcription followed by normal splicing of this gene will be (UTR: untranslated region)
    (A) 12.4kb   (B) 13.0kb   (C) 12.0kb   (D) 12.6kb
    (GATE 2006 Life Sciences)

4. Melting curve of two DNA specimens X and Y at the same pH and ionic strength have Tm values of 85°C and 80°C, respectively. This means that
    (A) The AT content of Y is higher than X
    (B) The GC content of Y is higher than X
    (C) The AT content is same in X and Y
    (D) The GC content of X is higher than Y   *(GATE 2006 Life Sciences)*

5. Plasmid is
   (A) Macromolecule involved in the protein synthesis
   (B) Circular piece of duplex DNA
   (C) A hybrid that is formed by joining pieces of DNA
   (D) Endogenous substance secreted by one type of cell
   (GATE 2003 Pharmaceutical Sciences)

6. When electroporation is used for introducing DNA into mammalian cells
   (P) A carrier for DNA is not required
   (Q) The lipid bilayer (membrane) interacts with an electric pulse to generate permeation sites
   (R) The viability of the cells becomes approximately zero percent
   (S) The first step involves absorption of DNA on the cell membrane

   (A) P, Q   (B) Q, R   (C) P, S   (D) Q, S
   (GATE 2006 Life Sciences)

7. G-CSF a myeloid growth factor
   (P) exhibits action similar to that of folic acid
   (Q) has a remarkable ability to mobilize haemopoetic stem cells
   (R) is activated by t-PA
   (S) activates phagocytic activity of mature neutrophils and prolongs their survival of circulation

   (A) Q, S   (B) P, Q   (C) Q, R   (D) R, S
   (GATE 2006 Pharmaceutical Sciences)

8. An example of haemopoetic growth factor is
   (A) Platelet derived growth factor   (B) Epidermal growth factor
   (C) Iron Dextran   (D) Erythropoietin
   (GATE 2004 Pharmaceutical Sciences)

9. The breakdown of fibrin is catalyzed by
   (A) Plasmin   (B) Renin   (C) Urokinase   (D) Ptylin
   (GATE 2004 Pharmaceutical Sciences)

10. Which of these is true about the discovery of HB antigen in the blood of people infected with Hepatitis-B?
    (A) It provided a basis for vaccine design
    (B) It indicated that specific vaccines cannot be designed for Hepatitis-B
    (C) It has not been of much significance
    (D) It indicated that Hepatitis-B is a viral disease
    (GATE 2004 Pharmaceutical Sciences)

11. DNA amplification by the polymerase chain reaction uses
    (A) *Thermus aquaticus* DNA polymerase (B) DNA topoisomerase
    (C) RNA polymerase   (D) DNAhelicase
                                (GATE 2004 Pharmaceutical Sciences)

B. Fill in the blanks
1. There are 20 amino acids and 64 codons, this implies that more than one triplet codes for an amino acid. This is referred to as _________________ of genetic code.

2. _________________ is the number of helical windings of DNA strands around each other while _________________ is the number of coilings of double helical axis over itself.

3. DNA can be bent at discrete sites. This feature of DNA is referred to as _________________ .

4. DNA polymerase-I is a trifunctional enzyme that can be separated by proteases into a 36kD small fragment having $5'\rightarrow3'$ exonuclease activity and 67kD large fragment called as _________________ having polymerase and $3'\rightarrow5'$ exonuclease activity.

5. The recognition and cleavage sequence of restriction endonucleases are_________________ i.e. they read the same in either direction.

6. In the bacteriophage DNA, there are short single stranded $5'$ projections of 12 nucleotides which are complementary in sequence and through which the bacteriophage DNA adopts a circular structure when injected into the host cell. These $5'$ terminal ends of bacteriophage DNA are referred to as_________________ .

7. The viruses that are used as vectors for introducing foreign genes into animal cells are _________________ and that used for insect cells are_________________ .

C. True or false
1. In eukaryotes DNA replication initiates at a single and unique origin of replication rather than at multiple random positions.

2. Under physiological conditions most of the DNA in prokaryotes and eukaryotes occurs in the classical Watson and Crick left handed B DNA form.

3. A relaxed DNA molecule is more compact than a supercoiled DNA molecule of the same length.

4. In DNA replication, DNA polymerase I removes the RNA primer by virtue of its $3' \rightarrow 5'$ exonuclease activity.

5. DNA Pol-I has proof reading activity by virtue of its $5' \rightarrow 3'$ exonuclease activity.

6. Gyrase catalyzes relaxation of supercoiled DNA, while topoisomerase I catalyzes the introduction of supercoils into DNA.

7. The staggered cuts made by a restriction endonuclease producing complementary single stranded ends, having affinity for each other, are known as blunt or square ends.

8. Recombinant human insulin is produced through genetic engineering by cloning and expressing genomic DNA encoding for human insulin in *Escherichia coli*.

9. DNA can be introduced into plant cells by bombardment with 1μm diameter chromium micro-projectiles coated with DNA.

ANSWERS TO OBJECTIVE TYPE QUESTIONS

A. Multiple choice questions

1. A	3. C	5. B	7. A	9. A	11. A
2. C	4. D	6. A	8. D	10. A	

B. Fill in the blanks

1. degeneracy	5. palindromic
2. twist number, writhe number	6. cos (cohesive) ends
3. kinking	7. retrovirus, baculovirus
4. Klenow	

C. True or false

1. False (right handed)	6. False
2. False	7. False (cohesive ends)
3. False	8. False (cDNA)
4. False ($5' \rightarrow 3'$ activity)	9. False (tungsten or gold particles)
5. False ($3' \rightarrow 5'$ activity)	

CHAPTER 3: ENZYMES AND ENZYME IMMOBILIZATION

ESSAY TYPE QUESTIONS

1. (a) Explain the regulation of enzyme activity.
    (b) Discuss inactivation of enzymes.
    (c) Explain the factors affecting enzyme activity.

2. What are immobilized enzymes? What are the various techniques for enzyme immobilization? Discuss the therapeutic applications of immobilization?

3. Write notes on
    (a) Enzyme electrodes
    (b) Isoenzymes
    (c) Enzyme metabolism

4. Define enzymes and discuss various factors affecting enzyme action with special reference to Michaelis-Menten equation.

5. Write notes on any three:
    (a) Isoenzymes;
    (b) Inactivation of enzymes;
    (c) ELISA
    (d) Enzyme immobilization on inorganic support

6. Define enzymes. Give classification of enzymes with examples. Write a detailed note on enzyme based sensors.

7. Write notes on:
    (a) Mechanism of enzyme inhibition
    (b) Michaelis-Menten equation
    (d) Antibiotic inactivation enzymes

8. Classify the enzymes. What is immobilization? Enumerate different methods of enzyme immobilization. Explain the regulation of enzyme activity.

9. Write notes on:
    (a) Enzyme electrodes
    (b) Isoenzymes
    (c) Enzyme metabolism

10. Write a note on the following
    (a) Allosteric enzymes   (b) hyaluronidase.   (c) streptokinase

OBJECTIVE TYPE QUESTIONS

A. Multiple choice questions

1. A term which describes a cofactor that is finally bound to an apoenzyme
    - (A) Holoenzyme
    - (B) Prosthetic group
    - (C) Coenzyme
    - (D) Transferase

    (GATE 2003 Pharmaceutical Sciences)

2. In the study of enzyme kinetics, V_{max} is said to be attained when
    - (A) There is an excess of free enzyme as compared to the substrate
    - (B) Virtually all of the enzyme is present as the enzyme-substrate complex and concentration of free enzyme is vanishingly small
    - (C) The maximum velocity of reaction in presence of low substrate concentration
    - (D) When the concentration of free enzyme equals that of enzyme-substrate complex

    (GATE 2005 Pharmaceutical Sciences)

3. The usefulness of 5-flurouracil as an antitumour can be attributed to one of the following mechanisms:
    - (A) It inhibits hypoxanthine-guanine phosphoribosyl transferase directly
    - (B) It is a prodrug that gets converted into fluro-2'-deoxy uridylic acid which is a suicide substrate for thymidilate synthase
    - (C) It gets incorporated into RNA leading to faulty transcription and translation into non-standard amino acid
    - (D) It gets converted into tetrafluro uridylate which inhibits purine nucleoside phosphorylase

    (GATE 2006 Pharmaceutical Sciences)

4. In competitive enzyme inhibition, the y-intercept of the plot between $1/V$ versus $1/[S]$
    - (A) is same in the presence and absence of inhibitor and slope is increased
    - (B) is decreased in the presence and absence of inhibitor and slope is increased
    - (C) is same in the presence and absence of inhibitor and slope is decreased
    - (D) is decreased in the presence and absence of inhibitor and slope is decreased

5. Enzyme immobilization is also known as
    - (A) Enzyme restriction
    - (B) Forced homing
    - (C) Entrapment
    - (D) A & B
    - (E) All of the above

6. Immobilization of enzymes
    (P) Increases the specificity of the enzyme for its reactants
    (Q) Facilitates reuse of the enzyme in batch reactions
    (R) Makes it unsuitable for its use in a continuous reactor system
    (S) Decreases the operational cost of the industrial process

    (A) Q, S   (B) Q, R   (C) R, S   (D) P, Q

    (GATE 2006 Life Sciences)

B. Fill in the blanks

1. The value of Michaelis–Menten constant, K_m, for most enzymes ranges between _________and________Moles.

2. A plot between reciprocal of reaction velocity, 1/V, versus reciprocal of substrate concentration, 1/[S], is called________________ plot which yields a straight line with an x-intercept of________________; y-intercept of________________ and slope of ________________.

3. In allosteric enzymes, the ________________ form has higher affinity for the substrate while the ________________ form has lower affinity for the substrate.

4. Plot between reaction velocity versus substrate concentration as displayed by allosteric enzymes is ________________ rather than hyperbolic as predicted by Michaelis-Menten model.

5. The advantages of using immobilized enzymes are three fold which include________________, ________________, and ________________.

C. True or false

1. Competitive inhibition, in contrast with non-competitive inhibition, cannot be overcome by increasing the substrate concentration.

2. Treatment of allosteric enzymes with mercurials, urea, proteolytic enzymes, high or low pH produces loss of catalytic activity with retention of feed back control.

3. According to Monod-Wyman-Changeu (MWC) model, the conformational change from T (Tense) to R (Relaxed) form occurs with conservation of symmetry i.e. all subunits in each state have the same conformation and the same intrinsic ligand affinity.

4. Lactate dehydrogenase occurs in animal tissue as four different isoenzymes.

5. Production of high-fructose syrup (used as a substitute for simple syrup) involves the conversion of glucose in part to fructose. This inter-conversion is brought about by the enzyme glucose isomerase used in its immobilized form.

ANSWERS TO OBJECTIVE TYPE QUESTIONS

A. Multiple choice questions
1. B   2. B   3. B   4. A   5. E   6. A

B. Fill in the blanks
1. 10^{-1} and 10^{-6}
2. Lineweaver- Burk plot; $-1/K_m$; $1/V_{max}$; K_m/V_{max}
3. R (relaxed); T (tense)
4. sigmoidal
5. reusability, purity, efficiency

C. True or false
1. False   2. False   3. True   4. False   5. True

CHAPTER 4: FERMENTATION

ESSAY TYPE QUESTIONS
1. Describe the production of Riboflavin by fermentation.
2. Describe the production of Ascorbic acid by fermentation
3. What is fermentation? Describe the production of penicillin by fermentation?
4. Describe in detail the fermentation process of streptomycin. Explain the recovery and purification process of streptomycin.
5. Explain the fermentation process of proteases and ascorbic acid.
6. Write notes on:
   (a) Fermentation process of penicillin
   (b) Fermentation process of Lysine
7. (a) Draw a labeled flow diagram for the fermentation based production of riboflavin
   (b) Discuss the newer processes for Vit-C preparation.

8. Write notes on:
   (a) Strains and culture media for penicillin fermentation
   (b) Ethanol-production

9. Discuss the fermentative production of:
   (a) Glutamic acid (b) Fungal amylases

OBJECTIVE TYPE QUESTIONS

A. Multiple choice questions

1. The stage of cell growth wherein cells adjust to their newer environment is
   (A) Latent phase (B) Exponential phase (C) Stationary phase   (D) Death phase

2. The aim in fermentation is to shorten the latent phase. This can be achieved by
   (P) Inoculating the cells growing in a seed culture medium having high nutrient concentration to a production culture medium having low nutrient concentration
   (Q) Inoculating a large volume (3%-10% of production medium) of actively growing cells
   (R) Inoculating the cells growing in a seed culture medium having low nutrient concentration to a production culture medium having high nutrient concentration
   (S) Inoculating a small volume (less than 3% of production medium) of actively growing cells

   (A) P, Q   (B) Q, R   (C) R, S   (D) S, P

3. The improvement of an industrially useful microbial strain can be achieved through
   (A) Somaclonal variation   (B) Mutation
   (C) Genetic recombination   (D) B, C   (E) All of the above

4. The foremost emphasis, in designing of a fermentor, is laid on the achievement and maintenance of
   (A) Appropriate rate of airflow   (B) Appropriate rate of heat transfer
   (C) Desired pH   (D) Aseptic conditions

5. Commercial production of citric acid is carried out by the microbial culture of
   (A) *Fusarium moniliformi*   (B) *Rhizopus*
   C) *Aspergillus niger*   (D) *Candida utilis*
   (GATE 2003 Pharmaceutical Sciences)

B. Fill in the blanks

1. In fermentative production of streptomycin, _______________ serves as the precursor of streptidine subunit of streptomycin molecule.

2. An example of nitrogen source most commonly employed for the fermentative production of penicillin is_______________.

3. ______________ is used as the carbon source for fermentative production of penicillin because it slowly hydrolyses into glucose and galactose over a prolonged period of time.

4. The strain of *Escherichia coli* (ATCC 12408) used for the fermentative production of L-lysine is a ______________ auxotroph.

C. True or false

1. *Penicillium notatum* is industrially employed for the fermentative production of penicillin.

2. The strain of *Micrococcus* No.541 used for the fermentative production of L-glutamic acid is α–keto glutaric acid auxotroph.

ANSWERS TO OBJECTIVE TYPE QUESTIONS

A. Multiple choice questions
1. A   2. A   3. E   4. D   5. C

B. Fill in the blanks
1. myoinositol   3. lactose
2. corn steep liquor   4. lysine

C. True or false
1. False   2. False

CHAPTER 5: PLANT CULTURE

ESSAY TYPE QUESTIONS

1. (a) Explain how tissue and cell cultures of plants are produced.
   (b) Explain how capacity of culture can be improved to produce and accumulate secondary metabolites.

2. Write note on phytopharmaceuticals.

3. Explain the applications of plant tissue culture in production of phytopharmaceuticals.

4. Discuss the principles involved in plant tissue culture. Explain the applications of plant tissue culture in production of phytopharmaceuticals.

5. Describe various types of plant cells and tissues which can be cultured on nutrient media. What is meant by cell suspension culture?

6. Write a note on:
   (a) Pharmaceuticals produced by plant cell culture

7. Explain tissue culture. Discuss the pharmaceutical applications of tissue culture.
       (Univ. of Raj. 2003)

OBJECTIVE TYPE QUESTIONS

A. Multiple choice questions

1. Formation of somatic embryos or embyogenic tissue directly from the explant without the formation of an intermediate callus phase is
   (A) Somatic embryogeneic response
   (B) Callus formation
   (C) Direct somatic embryogenesis
   (D) Premature germination
       (GATE 2006 Pharmaceutical Sciences)

2. Micropropagation of the plants is carried out through
   (A) Cross fertilization
   (B) Seed germination
   (C) Plant tissue culture
   (D) Grafting
       (GATE 2004 Pharmaceutical Sciences)

3. Which of the following direct transformation methods is applicable to intact plant tissues?
   (A) Calcium chloride and PEG-mediated transformation
   (B) Liposome-mediated transformation
   (C) Electroporation
   (D) Transformation using microprojectiles
       (GATE 2006 Life Sciences)

4. The most effective method for producing virus-free plant is
   (A) Root culture
   (B) Meristem culture
   (C) Somatic embryogenesis
   (D) Floriculture
       (GATE 2005 Pharmaceutical Sciences)

5. It is possible to initiate the development of complete plants from callus cell cultures by suitable manipulation of medium with respect to
   (A) Minerals   (B) Vitamins   (C) Carbohydrates   (D) Hormones
       (GATE 2003 Pharmaceutical Sciences)

Plant tissue culture of carrot is being developed in the laboratory on a semisolid White's medium. (For questions 6-8)

6. The micronutrient essential in the medium is
   (A) NaCl   (B) $CoCl_2$ (C) KCl   (D) $CaCl_2$

7. The pH of the medium is
   (A) 6.6   (B) 6.0   (C) 5.6   (D) 5.0

8. The tissue growth observed is
   (A) Undifferentiated cells suspended in the medium
   (B) Undifferentiated cells in clusters distributed in the medium
   (C) Differentiated mass of cells
   (D) Surface growth of undifferentiated mass of cells

   (GATE 2004 Pharmaceutical Sciences)

9. Fusion of protoplasts, generated from different species especially, is promoted by treatment with
   (A) Sodium nitrite   (B) Calcium chloride
   (C) PEG   (D) Electric shock   (E) All of the above

10. Biotechnologically, the culture that is most appropriate for the production of secondary plant metabolites is
    (A) Suspension culture   (B) Callus culture
    (C) Organ culture   (D) Meristem culture   (E) All of the above

B. Fill in the blanks

1. Full form of 2, 4-D is ________________.

2. In order to generate protoplasts, the cell wall of plant cells can be removed by treatment with lytic enzymes like ________________ and/or ________________.

3. The Ti and Ri plasmids are responsible for pathogenecity (tumour and root inducing respectively) and for the induction of synthesis of amino acid polymorphs or sugar derivatives known as ________________.

4. An example of a DNA plant virus that is most commonly used for cloning genes in plants is ________________.

5. Mitsui Petrochemicals, Japan, is producing ________________, a red colored dye, from cell cultures of ________________.

C. True or false

1. Auxins are derivatives of amino purine.

2. In clonal propagation, meristem (a tissue) is the explant (starting material).

3. Plant secondary metabolites include amino acids, vitamins etc.

4. Formation of variant clones, generated from protoplast culture or cell culture or any other plant culture method, expressing new characters absent in parent cells is called somaclonal variation.

ANSWERS TO OBJECTIVE TYPE QUESTIONS

A. Multiple choice questions

1. C	3. D	5. D	7. C	9. E
2. C	4. B	6. B	8. D	10. A

B. Fill in the blanks

1. 2, 4-dichlorophenoxy acetic acid
2. cellulase, pectinase
3. opines
4. cauliflower mosaic virus
5. *Lithospermum erythrorhizon*

C. True or false

1. False (aminopurine derivaties)
2. False (a single cell is explant)
3. False (these are primary metabolites)
4. True

CHAPTER 6: ANIMAL CULTURE

OBJECTIVE TYPE QUESTIONS

A. Multiple choice questions

1. A component that must necessarily be added in the minimal medium for animal culture is
    (A) 5%-20%v/v of homologous or heterologous serum
    (B) Essential amino acids
    (C) Glucose
    (D) Vitamin B complex

2. To promote attachment and spreading of anchorage-dependent animal cells, the surface of the culture vessel needs to be coated with
   (A) Trypsin   (B) Collagen   (C) Pronase   (D) Polyglycol
(GATE 2006 Life Sciences)

3. Which of the following monolayer culture systems have the highest surface area: medium ratio?
   (A) Roux bottle   (B) Spiracell roller bottle
   (C) Hollow fibres   (D) Plastic bag/ film
(GATE 2006 Life Sciences)

B. Fill in the blanks
1. The limit to which cells of a primary cell line divide (including by subculturing on fresh medium i.e. being 'passaged') before they die out is known as ________________.

2. HeLa stands for ________________.
3. RPMI stands for ________________.

C. True or false
1. In contrast to cells of primary cell line, cells of established cell lines that have undergone transformation become anchorage independent and can grow in suspension.

ANSWERS TO OBJECTIVE TYPE QUESTIONS

A. Multiple choice questions
1. A   2. B   3. C

B. Fill in the blanks
1. Hay Flick limit
2. Henrietta Lacks
3. Roosevelt Park Memorial Institute

C. True or false
1. True

CHAPTER 7: MONOCLONAL ANTIBODIES

ESSAY TYPE QUESTIONS

1. Discuss monoclonal antibodies and their applications in drug targeting.
2. Describe the method of preparation and applications of monoclonal antibodies.
3. Discuss the advantages and disadvantages of using monoclonal antibodies as cell surface markers.
4. Define monoclonal antibodies. Describe in detail the analytical and therapeutic applications of monoclonal antibodies.
5. Write notes on:
   (a) Limitations of monoclonal antibodies
6. Classify monoclonal antibodies. Discuss the characterization and storage of antibodies. Give the advantages of monoclonal antibodies.

OBJECTIVE TYPE QUESTIONS

A. Multiple choice questions

1. Which of the following two statements are true concerning the structure of antibodies
   (P) The length of variable region of heavy chain is same as the length of variable region of light chain
   (Q) The length of variable region of heavy chain is about thrice the length of variable region of light chain
   (R) The length of constant region of heavy chain is same as the length of constant region of light chain
   (S) The length of constant region of heavy chain is about thrice the length of constant region of light chain

   (A) P, R   (B) Q, R   (C) Q, S   (D) P, S

2. Monoclonal antibodies are: (tick the most appropriate option)
   (A) Tissue specific
   (B) Microbe specific
   (C) Antigen specific
   (D) Epitope specific

3. A hybrid derived from the fusion of a myeloma cell ($HGPRT^-$) with an antibody secreting B-lymphocyte ($HGPRT^+$) can be selected to produce monoclonal antibody by growing in a medium containing
   (A) thiamine, hypoxanthine, aminopterine
   (B) thymidine, histidine, aminopterine
   (C) uridine, hypoxanthine, aminopterine
   (D) thymidine, hypoxanthine, aminopterine   *(GATE 2006 Life Sciences)*

4. HAT medium used for the selection of hybridoma cells was devised by
    (A) Georges Kohler
    (B) Cesar Milstein
    (C) Fred Grifth
    (D) Littlefield

5. Diversity in antibody is brought about by
    (A) Post-translational modifications
    (B) Gene rearrangements
    (C) Usage of special genetic codes
    (D) Multiple mutations in the polypeptide   *(GATE 2005 Pharmaceutical Sciences)*

6. Monoclonal antibodies in contrast to polyclonal antibodies are:
    (P) Single molecular species
    (Q) Homogeneous since they are produced by population of identical cells i.e. a clone
    (R) Heterogeneous since they are produced by different population of antibody producing cells
    (S) Different in their precise specificity and affinity for the antigen

    (A) P, Q   (B) P, R   (C) P, S   (D) Q, S

B. Fill in the blanks

1. The antigen binds at a particular site on the antibody called as the hypervariable region. This is also known as________________ because it determines antibody specificity.

2. The mammalian cells synthesize nucleotides by two different pathways which include________________ and ________________.

3. The injection of murine monoclonal antibodies into human patients results in an immune response which leads to the development of ________________.

C. True or false

1. HGPRT deficient myeloma cells do not survive when cultured in HAT medium.

ANSWERS TO OBJECTIVE TYPE QUESTIONS

A. Multiple choice questions

1. D   2. D   3. D   4. D   5. B   6. A

B. Fill in the blanks
1. complementarity determining regions
2. *de novo* pathway; salvage pathway
3. human anti murine antibodies (HAMA)

C. True or false
1. True

CHAPTER 8: MICROBIAL TRANSFORMATION

ESSAY TYPE QUESTIONS
1. Write a note on:
    (a) Biotransformation of antitumour drugs

OBJECTIVE TYPE QUESTIONS

A. Multiple choice questions
1. Two important advantages of using micro-organisms for biotransformation in drug synthesis are
    (P) having been produced from micro-organisms, they are certain to have antibacterial properties
    (Q) they are abundant in nature and hence reduce the processing cost significantly
    (R) they produce the specific stereoisomer only
    (S) they are highly selective and yield products with high purity

    (A) P, Q   (B) Q, R   (C) P, S   (D) R, S
    (GATE 2006 Pharmaceutical Sciences)

ANSWERS TO OBJECTIVE TYPE QUESTIONS

A. Multiple choice questions
1. D

CHAPTER 9: CELL IMMOBILIZATION

OBJECTIVE TYPE QUESTIONS

A. Multiple choice questions
1. Which of the following statements are INCORRECT about immobilized plant cell cultures?
    (A) It is possible to use high cell densities
    (B) Cells remain active for long periods
    (C) Cell products or inhibitors can be removed easily
    (D) It provides low shear resistance to cells   *(GATE 2006 Life Sciences)*

ANSWERS TO OBJECTIVE TYPE QUESTIONS

A. Multiple choice questions
1. D

CHAPTER 10: HALOMETABOLITES

ESSAY TYPE QUESTIONS
1. Write notes on:
   (a) Halometabolites

OBJECTIVE TYPE QUESTIONS

A. Multiple choice questions
1. The major source of halometabolites is
   (A) Marine algae (B) Terrestrial plants (C) Mammals   (D) Fungi

2. Characteristics of bromoperoxidase, a haloperoxidase, include
   (A) It oxidizes bromide ion
   (B) It has an optimum pH of 5-7 for halogenating activity
   (C) It is a haem containing enzyme
   (D) It lacks specificity of action
   (E) All of the above

3. Applications of halometabolites include
   (A) Role in defense mechanism such as antimicrobials, antifeedants, antinsecticides
   (B) Biological activities such as α-, β-adrenergic activity, cholinergic activity, stimulation of DNA synthesis, bronchodilation, muscle relaxation
   (C) Biosynthesis e.g. formation of terpenoids, eonic acid, lactones, pyrones
   (D) All of the above

B. Fill in the blanks
1. One of the mechanisms explaining halometabolite biosynthesis is based on the fact that a free ________________ acid is primarily formed by enzymatic catalysis which secondarily reacts with a broad range of substances to yield the respective halometabolite.

ANSWERS TO OBJECTIVE TYPE QUESTIONS

A. Multiple choice questions
1. A   2. E   3. E

B. Fill in the blanks
1. Hypophalous acid

CHAPTER 11: BIOMASS: ENERGY PRODUCTION

ESSAY TYPE QUESTIONS
1. Write notes on:
   (a) Importance of Bioenergy
   (b) Energy production by biomass

2. Write notes on:
   (a) Hydrocarbons as source of energy
   (b) Importance of Bio-mass.

3. Discuss production of energy Bio Mass.

4. Write short notes on
   (a) Bio energy

5. Write short notes on
   (a) Bio gas production

OBJECTIVE TYPE QUESTIONS

A. Multiple choice questions
1. Biohydrogen can be produced by
   (A) Electrolysis of water by employing photosynthetic machinery of green plants
   (B) Employing hydrogenase enzyme of microbes such as *Chlamydomonas* etc.
   (C) Employing nitrogenase enzyme of cyanobacteria such as *Anabaena* etc.
   (D) Reforming of fossil fuels such as natural gas
   (E) All of the above

B. Fill in the blanks

1. Pyrolysis is destructive distillation of organic matter carried out in the absence of
   _________________ gas and at a high temperature of _________________ °C.

2. Brazil is a country where all cars run either on _________________ or its mixture
   with petrol.

3. Production of biogas from cow dung can be enhanced by addition of green algae
   such as _________________ .

ANSWERS TO OBJECTIVE TYPE QUESTIONS

A. Multiple choice questions
1. E

B. Fill in the blanks
1. Oxygen; 200°C-500°C   2. Alcohol   3. *Zygogonium* sp.

SOME USEFUL WEBSITES

Website Address	Provides Access To/ Information About
Genetics	
www.dnaftb.org/dnaftb	A-Z of DNA
www.biospace.com	Genetic engineering
www.biotech-info.net	Genetic engineering, ecology
en.wikipedia.org/wiki/Genetic_engineering	Genetic engineering
www.ornl.gov	Human Genome Project
Bioinformatics	
www.genome.ad.jp	Bioinformatics
www.bioinformatics.org	Bioinformatics
www.imtech.res.in/bic/	Bioinformatics
www.123genomics.com	Genomics
www.ncbi.nlm.nih.gov	National Center for Biotechnology Information –Genomics- nucleic acid sequence database
www.ebi.ac.uk	European Molecular Biology Laboratory - nucleic acid sequence database
www.ddbj.nig.ac.jp	DNA databank of Japan- nucleic acid sequence database
www.pir.georgetown.edu	Protein Information Resource- protein sequence database
www.us.expasy.org	Swiss-Prot- protein sequence database
www.rcsb.org/pdb	The RSCB Protein Data Bank- protein sequence database
www.ndbserver.rutgers.edu	Nucleic acid structure database
www.cubic.bioc.columbia.edu/predictprotein	Protein Structure Prediction and Sequence Analysis
www.cstl.nist.gov/div831/strbase	Short Tandem Repeat DNA Internet DataBase
dbtindia.nic.in/bits/dic.html	List of Bioinformatics distributed information centers in India
dbtindia.nic.in/bits/subdic.html	List of Bioinformatics distributed information subcenters in India
Gene therapy	
www4.od.nih.gov/oba/Rdna.html	Gene therapy
www.dnapolicy.org/ gene therapy	Gene therapy
www.nature.com/gt/index.html	Gene therapy, stem cells

Animal Culture

www.med.umich.edu/tamc/links.html	Transgenic animals
www.agresearch.co.nz/scied/search/biotech/gene_gmomaking_animal.htm	Transgenic animal

Enzymes

www.fhsu.edu/chemistry/	Enzyme kinetics
www.web.indstate.edu/	Enzyme kinetics
www.rohmhass.com	Enzyme immobilization
www.ornl.gov	Biosensors

Fermentation

www.en.wikipedia.org/wiki/fermentation	Fermentation

Antibodies

www.antibodyresource.com	Antibodies
www.antibody.bath.ac.uk	Antibody structure

Plant Culture

www.agbios.com/default.asp	Agricultural Biotechnology Strategies, GMO
www.aggie-horticulture.tamu.edu/tisscult	Plant tissue culture

Products

www.lilydiabetes.com/products/humulin	Humulin, recombinant human insulin
www.gene.com/products/activase	Activase, rh tissue plasminogen activator
www.wwwext.amgen.com/products/	Epogen, apoetin alpha, rh erythropoietin, the red cell factor
www.gene.com/products/protropin	Protropin, met rh growth hormone
www.humatrope.com	Humatrope, rh growth hormone
www.gene.com/products/pulmozyme	Pulmozyme, rh DNase, hydrolyzes DNA in sputum of patients with cystic fibrosis
www.rocheusa.com/products/roferon	Roferone-A, rh Interferon α-2a
http://www.hivandhepatitis.com/index.html	Intron-A, rh Interferon α-2b
http://www.vaccinetruth.org/index.htm	Recombivax-HB, a non-infectious subunit viral vaccine derived from hepatitis B surface antigen (HBsAg) produced in yeast cells
http://rxusainternational.com/	Various biotech based products

Journals

www.nature.com/nbt	Monthly journal dealing with all aspects of biotechnology
www.biomednet.com/home	Online publishing, databases, reviews
www.sciencekomm.at	Journal database, links
www.spectroscopynow.com/Spy/basehtml	Online publication

www.cato.com/biotech	Journals
www.ncbinlm.nih.gov/Entrez	Bibliographic database
www.bmn.com	Bibliographic database
www.cato.com/biotech	Virtual library

News

www.bioworld.com	Biotech newspaper
www.genwirl.com	Genetic engineering news
www.biospace.com	Biotech business
www.dnavaccine.com	Biotechnology, news
www.genomeweb.com	Biotech News
www.webcom.com/pgi/	Biomedical news
www.biotaq.com	Biotech News, press info
www.biotechresearch.biz	Biotech News
http://search.1millionpapers.com	Biotech News
www.genengnews.com	Biotech News

Search Engines

| www.biotechfind.com | Biotechnology, search engine |
| www.scirus.com | Scirus- Scientific search engine |

Miscellaneous

www.pharma.org/charts/b_coo.html	Biotechnology medicines in development
www.academicinfo.net	Academic Info- Education links
www.biointeractive.org	Educational multimedia
www.sciweb.com	Molecular biology links, news, jobs
www.glossarist.com	Definitions
www.everythingbio.com	Glossary/ definitions
www.bio.com	Companies involved in biopharmaceuticals, biotechnology, diagnostics and pharmaceuticals, as well as product and equipment suppliers
www.nal.usda.gov/bic/Education_res	General education material etc.

Eminent scientists

Sir Alexander Fleming 1881-1955

Friedrich Miescher

Dr. Hargovind Khorana   (1922-)

Matthew Meselson

Franklin Stahl

Kohler, Georges (1946-1995)

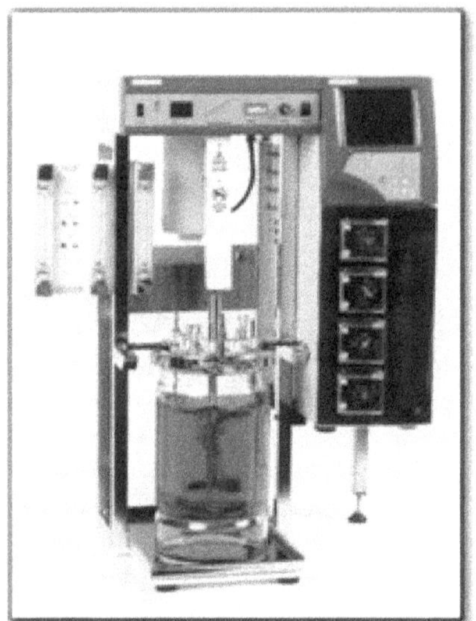

A typical laboratory scale fermentor with inlets for various provisions

A laboratory scale airlift fermentor

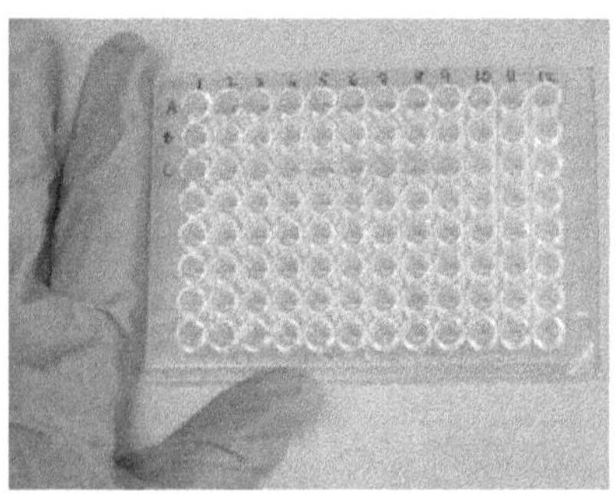

A 96-well microtiter plate used for ELISA

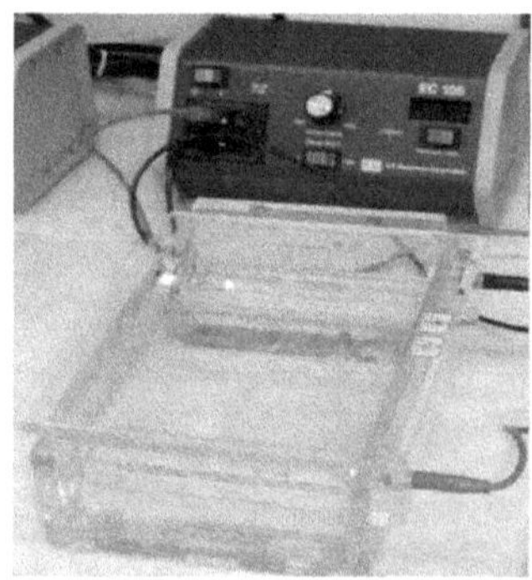

Gel electrophoresis kit

Particle Delivery System

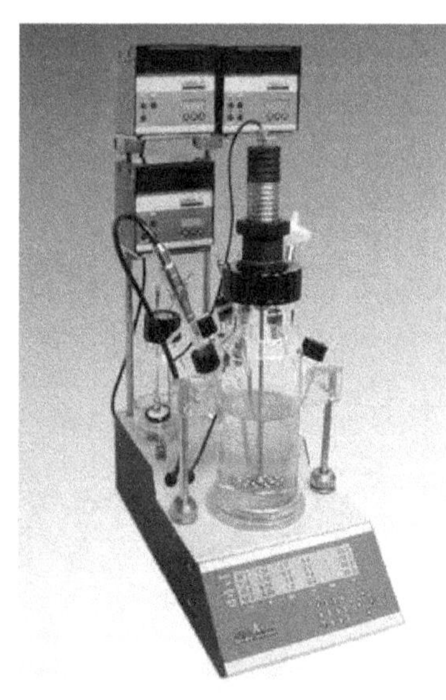

A laboratory scale fermentor

Genetic engineering

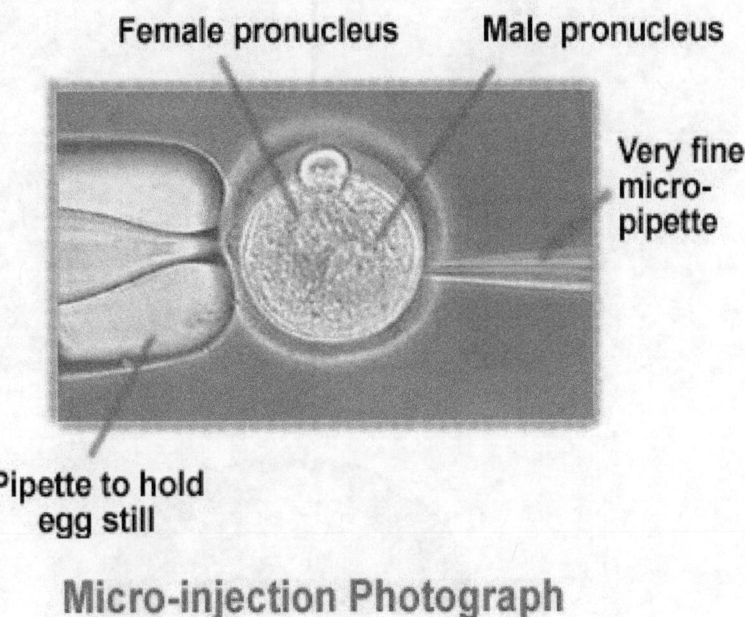

Micro-injection Photograph

Fermentation

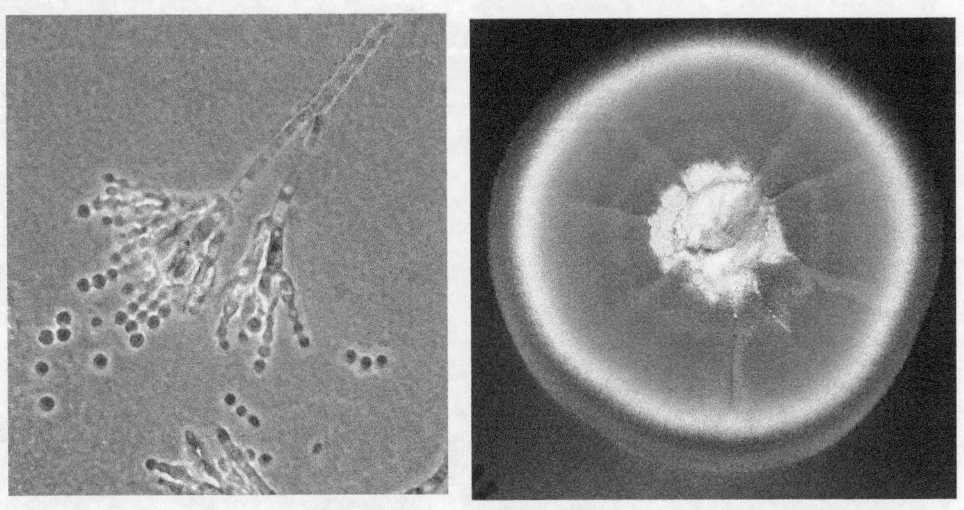

Penicillium chrysogenum

Crown gall

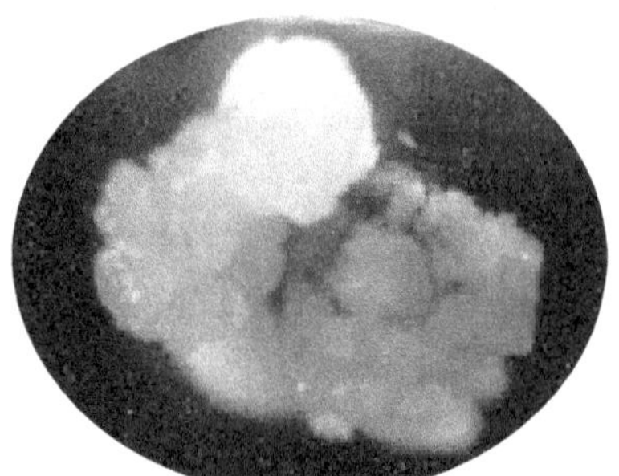

Callus culture

Green house

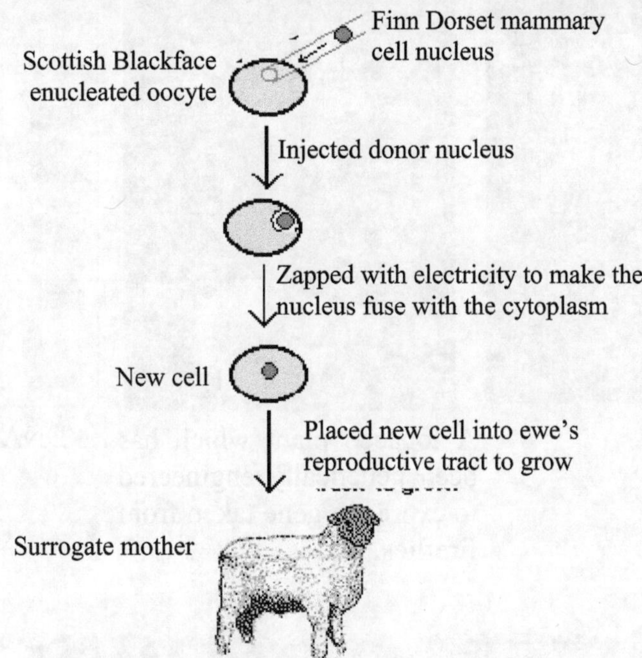

This is how Bill Ritchie (a technician working with Dr Campbell) made Dolly.

Dolly, the cloned sheep, first ever mammal to have be cloned.

Dolly

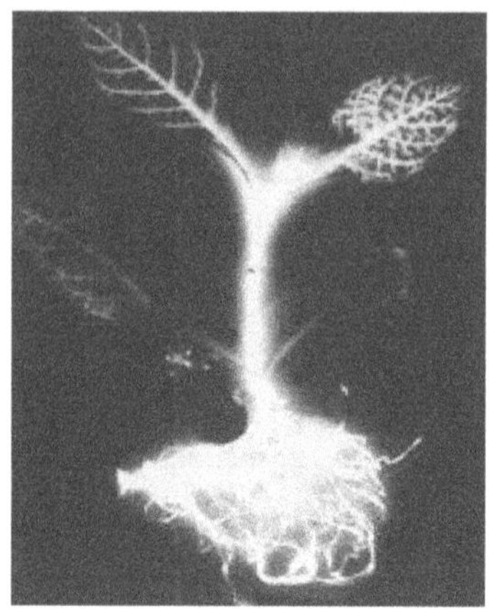

A tobacco plant which has been genetically engineered to express a gene taken from fireflies.

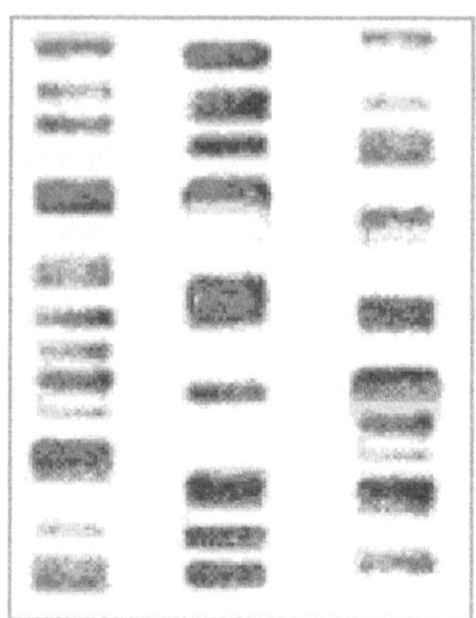

DNA fingerprint

Humulin Pen

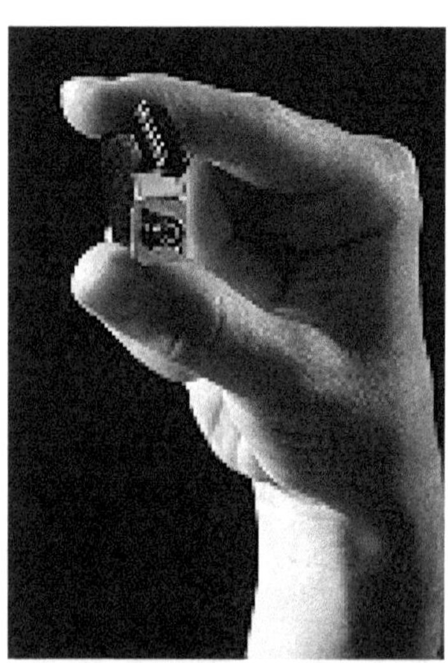

The infrared microspectrometer developed at ORNL can be used for blood chemistry analysis, gasoline octane analysis, environmental monitoring, industrial process control, aircraft corrosion monitoring, and detection of chemical warfare agents.

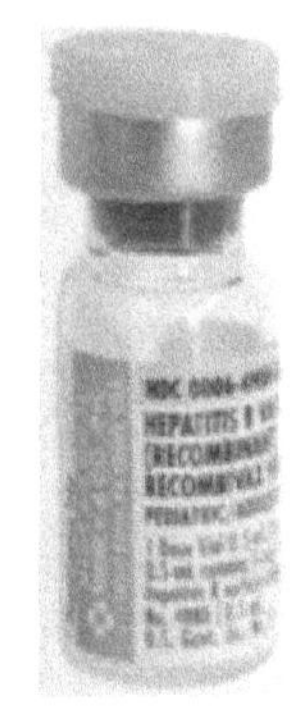

Recombivax HB

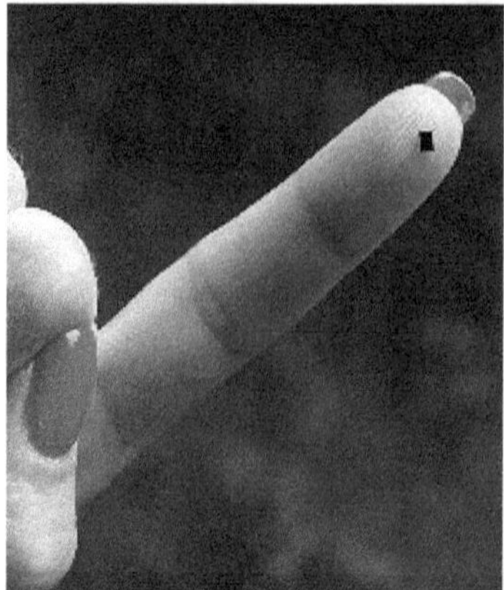

This "medical telesensor" chip on a fingertip can measure and transmit body temperature.

...IT IS ONLY THE BEGINNING

Index